PRODUCTIVE WATER POINTS
IN DRYLAND AREAS

PRODUCTIVE WATER POINTS IN DRYLAND AREAS

Guidelines on integrated planning for rural water supply

CHRIS LOVELL

ITDG
PUBLISHING

in association with
the Centre for Ecology and Hydrology

Published by ITDG Publishing
103-105 Southampton Row, London WC1B 4HL, UK

© Centre for Ecology and Hydrology 2000
First published in 2000

ISBN 1 85339 516 1

A catalogue record for this book is available from the British Library

ITDG Publishing is the publishing arm of the Intermediate Technology
Development Group. Our mission is to build the skills and capacity of people in
developing countries through the dissemination of information in al forms,
enabling them to improve the quality of their lives and that of future generations.

Edited and typeset by the Centre for Ecology and Hydrology
Printed in Great Britain by Short Run Press Limited Exeter

Contents

vi

Preface

LACK OF WATER IS preventing many household and community-based activities for millions of people living in dryland areas of developing countries. When water becomes available it is put to a wide variety of uses: drinking, washing, food processing, beer brewing, brick-making, vegetable growing, fruit orchards, livestock feedlots, small-scale dairies, fish farming etc. Many of these water-related activities have a high economic value. They can play an important role in household income and livelihood strategies, and through diversification can avoid over-reliance on single production activities such as rainfed cropping of marginal lands. However, the diverse range of production strategies that can be associated with a water point have not formally been promoted. Rural water supply programmes have tended to focus on only two social aspects: improved access to domestic supply and improved sanitation. Less attention has been paid to exactly how a community would prefer to use the water to develop its own livelihoods. This is due in part to the difficulties of abstracting sufficient reliable groundwater in dryland areas, and in part to a misunderstanding of why wells and boreholes fail, which leads to a general belief that abstraction should be limited to domestic supply to conserve the resource. Lack of water is also exacerbated in many instances because communal water points lie broken and local people have neither the capacity nor the incentive to repair what has become, in effect, a state-owned common property resource.

Many of the world's dryland areas are underlain by crystalline basement rocks, the most common being gneisses and granites. The aquifers in these 'hard rocks' are important because of their widespread extent and because for many people there is no readily available alternative source of water supply. Traditionally they have been exploited by shallow hand-dug wells, more recently by deep boreholes, but in many cases the yields are low and many of these water points fail during drought. In the early 1980s, a research programme began in southern Africa to study the feasibility of developing hard rock aquifers to support income-generating activities such as small-scale irrigation, and to assess the potential of productive water points as an initial step towards establishing sustainable community-based management of natural resources in drought-prone areas.

Productive water points in this context are generally community-managed, designed and implemented as part of rural water supply programmes to provide water surplus to domestic needs which may be used for economically

vii

productive purposes. Importantly, they are implemented in a manner that empowers the local user group to own and maintain the resource. Income from production provides both the incentive and the cash needed for local maintenance and relieves pressure on the state where this is now failing to maintain rural water supplies. In the longer term, reinvestment of this income in the local production system, combined with experience of successful collective action, creates a range of much wider benefits to the local economy and environment through diversification of livelihood strategies and intensification of the dryland farming system.

The research in southern Africa has shed light on why conventional wells and boreholes fail, on the potential of the groundwater resource to support production through improved siting and selection of appropriate well designs, and on the impacts that productive water points can have on community resource management and livelihood strategies. The aim of this book is to present the findings of this research in a practical manner, to encourage planners and practitioners in rural water supply to consider developing productive water points in drought-prone areas, and to provide the information they need to follow this through. One of the main constraints to adopting this strategy has been that experience and information on it was limited to a few countries. I hope that this book will partly overcome that.

As the book is intended to be practical in approach and not an academic treatise, few references are given in the text but are listed after each chapter to help the reader understand the origin of the material presented and to follow it up if desired. Particular attention is given to answering the questions most commonly asked about productive water points. Although the book draws extensively on design and construction experience gained by staff on projects in southern Africa (notably Zambia and Zimbabwe) the guidelines are considered relevant to many dryland areas. Although the focus is on groundwater, the book also draws on experience gained by staff on projects to develop small dams, and in particular, the institutional arrangements so critical to management of community-based projects. In this regard the guidelines are relevant to ground and surface water schemes alike.

The planning, siting, selection, implementation and management of productive water points in drought-prone areas are considered, and the critical issue of working in partnership with the community occupies an important place. The book aims to provide decision-making support to help choose and implement the most appropriate and cost-effective option in any area. Chapter 2 provides an overview of the groundwater resource and why wells and boreholes fail. Alternative well designs are described and classified to enable siting and selection of the most appropriate technology in different ground conditions. Chapter 3 considers the economic costs and benefits of the various ground and surface water options available, and compares them with standard domestic water points. Chapters 4 and 5 provide flow charts, decision trees and detailed information on how to make more effective use of existing water points, and on siting, selecting and constructing new water points of appropriate design. Chapter 6 focuses on a step-wise approach to working in partnership with the community, placing emphasis on local ownership and participation throughout all stages from planning and design to construction, operation and maintenance. Community-based projects do raise special management problems, especially where a common property resource such as water assumes an economic value through production. Issues emerging at productive water points in southern Africa are considered in Chapter 7, as well as environmental issues such as groundwater depletion now occurring in some parts of Asia. Chapter 8 concludes by anticipating the future of rural water supply in dryland areas and the broader institutional arrangements that will be needed in an environment of increasing population pressure.

Given the range of physical and social settings found in the world's dryland areas, a single book cannot offer all the information needed to develop productive water points. This book is a synthesis of lessons learnt in a decade developing productive water points in southern Africa. The approach outlined has been found helpful in collaborating with rural communities in communally-managed dryland areas, and successful in abstracting groundwater for production from hard rock aquifers typical of the region. Where similar conditions exist in the world, these are outlined. I hope that parts of the approach are helpful to you in your particular setting. I wish you luck and would welcome information on your experiences in other dryland areas, as well as comments, corrections and additions to the materials presented in this book.

Chris Lovell
Centre for Ecology and Hydrology
Wallingford, UK

Acknowledgements

This book is an output from various projects funded by the UK Department for International Development (DFID) for the benefit of developing countries (projects R7131, R5846, 5849, 5851, 5551, 6085, 'Small-scale irrigation using collector wells - Zimbabwe', and 'Potential for productive groundwater development in the Southern Province of Zambia'). The views expressed are not necessarily those of the DFID.

I wish to acknowledge the data and assistance received from many organisations involved through these and other projects on the design and implementation of productive water points. Invaluable assistance has been received from the British Geological Survey (BGS), the Zimbabwe Departments of Research and Specialist Services (R&SS), Agricultural Extension (Agritex), Water Development (DWD) and District Development (DDF), Plan International in Zimbabwe, and CARE International in Zimbabwe and Zambia, who collaborated with the former Institute of Hydrology during various stages of this work. However, any errors are entirely my own, and should not be attributed in any manner to these organisations.

On a personal level, I would like to thank the following:
Cath Abbott, David Adriance, Jim Alexander, Dominic Banga, John Barker, Clare Barrington, John Bell, Eric Bemont, Mike Brown, John Butterworth, Jeremy Cain, Bruce Campbell, John Carter, Richard Carter, Eric Chidenga, Francis Chishiri, R Chitsiko, Timothy Chiunye, Mike Clowes, Diana Conyers, Ian Curtis, Jeff Davies, Dale Dore, Mike Edwards, Mr Guni, Amanda Hammar, John Hansell, Robin Herbert, Martin Hodnett, Norbert Honigmann, Elizabeth Jones, Emlyn Jones, Sylvester Kalonge, Alex Kremer, Chris Lewcock, Felix Lombe, Eliah Mafunga, Andrew Mahlekete, Jo Makadho, Daniel Maringa, Edward Mazhangara, Isiah Mharapara, James Milford, Chris Mhlanga, Steven Mhlauri, Peter Morgan, Shelton Msika, Francis Mugabe, Osmond Mugweni, Monica Murata, Peter Musanhu, Zvenyika Mutema, Godwin and Miriam Mtetwa, Elijah Mwashayeni, Fanuel Nangati, Peter Ngoreto, Tony Peacock, George Nhunhama, Phibion Nyamudeza, Mthubani Nzima, Eric Petersen, Caroline Plastow, David Price, Peter Rastall, Hugh Rendall, Karen Sage, Andrew Semple, Mercus Sharpe, Lester Simmonds, Juliet Sithole, Kelly Stevenson, Sam Sunguro, John Tabene, Andy Tainsh, John Talbot, Duncan Thompson, Wilfred Tichagwa, George Tobiawa, Paul van Beers, Anthony Waterkeyn, Dominic Waughray, Irene Welch, Alistair Wray, Peter Wurzel, Sibanda Zanamwe, friends at Chiredzi Research Station, and the communities of Romwe, Muzondidya, Gokota, Dekeza, Nemauka, Mawadze, Matedze, Machoka and Masekesa for their various contributions to the research, and to John Bromley, John Chilton, Terence Dube, John Gash, Henry Gunston, Celia Kirby, Caro Lovell, David Macdonald, Patrick Moriarty and Caroline Sullivan for their support and help in producing the book.

Finally, very special thanks are due to Charles Batchelor who initiated and contributed so much to the work presented here and without whom this book would not have become possible.

1   Productive water points in dryland areas

What is a productive water point?

A PRODUCTIVE WATER POINT is a public water point, designed and implemented to provide water surplus to domestic needs which can be used for economically productive purposes. It can form part of a strategy to alleviate poverty and improve quality of life for communities in dry areas threatened by water shortage and livelihood insecurity.

A productive water point will typically serve upwards of 20 families, in an area where existing water points are scarce, low yielding or unreliable and where setting up private schemes for all members of the community is not viable (Box 1). It may use water from below the ground (groundwater) or surface water from a small reservoir or 'dam'. This book focuses on groundwater, and on the potential to develop hard rock aquifers in dryland areas, but the sections on community planning and implementation, and on community ownership and responsibility for operation and maintenance, apply equally to surface water-based schemes.

The potential income-generating activities that people can adopt at productive water points are numerous and wide-ranging. This book focuses on small-scale irrigation (particularly of vegetable gardens and fruit orchards) as this is an important drought coping strategy in many dryland areas. It could equally be fodder production for livestock feeding centres or dairy units or beer brewing or brick-making, depending on community priorities and traditions.

The term 'small-scale' is used here to define the situation where large numbers of individuals or member households (sometimes termed 'small-holders' or 'small-farmers') group together on large- (or small-)scale schemes. The important distinction is neither the scale of these schemes, nor the scale of the members' plots within them, but the nature of the scheme management. The terms 'communally-managed' and 'farmer-managed' are used here to emphasize that, to be successful, schemes should be structured for independent management by the users, as opposed to schemes requiring management by external agencies. With this approach, the scale of the scheme (and the choice of infrastructure) depends on what is manageable for the users, instead of the infrastructure dictating the management structure.

1

Box 1: Family wells and community wells

Productive water points can be either privately owned and operated, developed by an individual family primarily for their own use, or be community-based, typically serving upwards of 20 families, usually in areas where multiple water sources such as family wells are hard to develop or are low yielding and unreliable. When planning productive water points, low-cost family wells will be the first and logical choice in those areas where they can readily supply sufficient reliable water. The advantages of family wells include:

○ the relative ease with which they can be set up and managed;
○ decision-making by the single household avoids leadership disputes inevitable with community schemes;
○ maintenance by the single household does not rely on collective action;
○ if sited close to the homestead, they can have a direct impact on health, which has been found to rely on increasing the quantity of water available less than three minutes travel time from the home.

However, family wells in dryland areas are often low yielding and unreliable, especially during drought. In these areas, the under-utilized potential of groundwater, the equity of water supply and income generation within the community, and the drought resilience of those at risk, can best be realized by implementing low-cost family wells *and* more expensive but reliable community wells. The advantages of community wells include:

○ can be sited in optimum groundwater locations, thereby making effective use of the groundwater resource and ensuring that the most reliable supply is available to all during drought;
○ can reduce competition for water resources between private farmers and reduce the problems of groundwater management that arise with development of multiple private wells;
○ can build social capital, providing communities with the experience of successful collective action and the confidence and institutional structures required to tackle other natural resource management problems at the village level.

This book is intended to promote a judicious mix of both private and community-based water points.

Where will productive water points be useful?

Productive water points are needed most, and will be most successful, where:

○ people are relatively poor
○ rainfall is low and unreliable and drought is recurrent
○ water shortage is a key constraint to development
○ conventional wells and boreholes are difficult to site and often fail

- population densities are relatively high
- livelihoods are insecure
- there is increasing competition for land
- previously extensive farming systems are being forced to intensify
- many families, particularly young, are seeking alternative sources of livelihood.

These conditions are most common in dryland areas of the world underlain by crystalline basement rocks (Figure 1.1; Box 2) and particularly in communally managed dryland areas where land and water are managed as common property resources (Box 3). Dryland in this context is defined as semi-arid, having mean annual rainfall between 200 and 750 mm that is generally unreliable, highly variable, and typically falls for only a few months of each year. Livelihood insecurity is a feature of these low potential lands. Drought is an ever-present threat that can cause widespread crop failure and human suffering. Even in years of good rainfall, high intensity storms, soil erosion, water-logging, nutrient leaching, and the prevalence of pests and diseases, can reduce crop yields.

Given these adversities, everything in the culture will tend to focus on the need to manage risk, and most people will depend on a mix of activities to meet livelihood needs. Off-farm remittances from family members working away from home will often be important, as will extensive rainfed crop production, livestock, cash from the sale of local crop and woodland products (such as beer, honey, caterpillars, mushrooms, firewood, building materials) and cash from other small-scale enterprises such as irrigation of vegetables and fruit trees, gold panning and craftwork.

Have productive water points been developed before?

Productive water points have been developed before in some dryland areas where local communities and development agencies have recognized favourable groundwater conditions (or dam sites) and have exploited a readily available resource. *Fadama* irrigation from shallow 'self-help' wells and hand-drilled tubewells on the alluvial aquifers of northern Nigeria are examples. This book focuses elsewhere, on the major parts of the Earth's surface underlain by basement or 'hard rock' aquifers where the wide-scale development of productive water points as an integral part of national development programmes has been constrained by:
- the difficulties of abstracting groundwater and the corresponding shortage of reliable sources;
- scepticism among planners about the potential importance of developing informal, small-scale farmer-managed irrigation schemes and community-based enterprises;
- misgivings based on the poor performance of larger, formal irrigation schemes (Box 4).

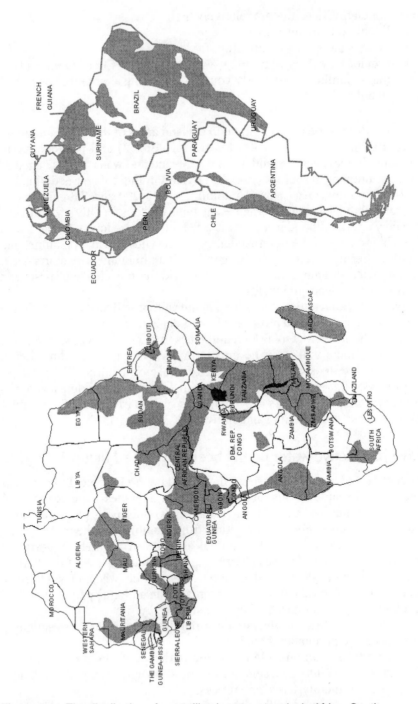

Figure 1.1 The distribution of crystalline basement rocks in Africa, South America and Asia (redrawn from Uhl and Atobrah, 1987)

4

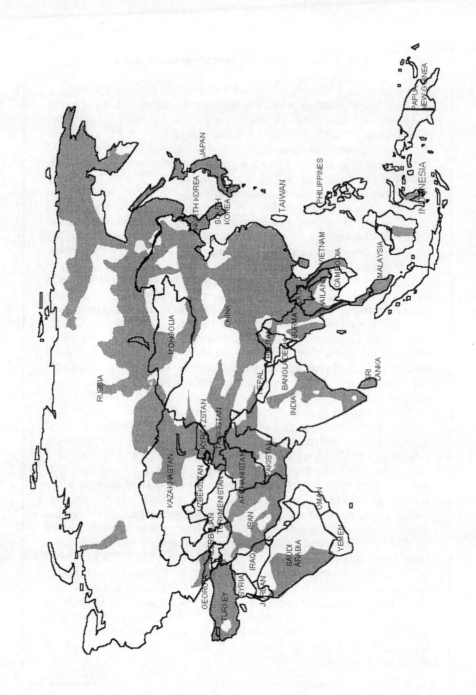

Box 2: Crystalline basement areas of the world

Over major parts of the earth's surface, the basement of continents outcrop and lie close to the surface in vast shields of igneous and metamorphic rocks. The most common of these 'hard rocks' are gneisses and granites, crystalline limestone, quartzites, schists and dolerites. Large parts of this geological domain are located in dryland regions which are among the least developed in the world (Figure 1.1). From a social and environmental perspective, these are among the most fragile parts of the world, with increasing populations having prevailing low per capita income. This is especially true of parts of Brazil, peninsular India, Sri Lanka, Korea, China, the Red Sea region, west Africa from Mauritania and Senegal to Burkina Faso and Cameroon, and extensive parts of central, eastern and southern Africa. The populations concerned are at least in the range of 30 million in Latin America, and several hundreds of millions in Africa and Asia.

One of the primary constraints on economic and social development in these regions is the difficulty encountered in developing reliable water supplies. In tropical arid regions, surface water is not available on a permanent basis, and it is often contaminated with waterborne pathogens. As a result, groundwater is, generally, the only permanent and safe source of water. However, the search for groundwater and its development in these regions raises a number of problems which until recently were considered almost impossible to solve. Hard rock aquifers are complex in occurrence and spatially highly variable. They have virtually no primary porosity, unlike sandstones and other sedimentary rocks. Instead, they have a secondary porosity due to fracturing and weathering which permits limited flow and storage of water. In tropical and sub-tropical regions, the weathered and often clayey overburden (regolith) provides the main groundwater storage.

Boreholes may be completed in the underlying fractured bedrock but yields are low unless interconnected fractures are intercepted that draw on water stored in the regolith. Traditional dug wells, on the other hand, are completed in the regolith, but again yields are low where siting fails to identify sufficient depth of saturated regolith and where low permeability quickly causes localized drawdown of the water table around the well on pumping. Success rates, commonly defined as a sustained handpump yield of $0.3\ \mathrm{l\ s^{-1}}$ (5–10 $\mathrm{m^3/day}$), average some 60 per cent in the majority of dryland areas. There is thus the major cost of substantial numbers of failed wells and boreholes, which most developing countries could well wish to avoid. In addition, an average of some 60 per cent implies a much lower rate in the more difficult areas where poor people often live. In such locations, and following one or more borehole failures, the community is effectively being 'written off' by the current approach.

Rural water supply policy has tended to focus on only two social aspects, namely improved access to domestic supply and improved sanitation. Less attention has been paid to investigating exactly what *type* of water supply

Box 3: Communally managed dryland areas

The communally managed areas of Zimbabwe typify the problems now facing people and the environment in many low rainfall parts of Africa.

In the nineteenth century the limited scale of subsistence agriculture imposed little stress on the environment. Although periodic drought had an impact, the effects were largely ameliorated by migration and rotational methods of resource utilization. Today, however, these traditional drought coping strategies are inappropriate as a result of population increase, competition for land, and changes in land tenure brought about by colonialization and land apportionment. Fifty-seven per cent of people live in 'communal lands' designated in the colonial era. All land is owned *de jure* by the sate but, *de facto*, cropping land is owned by families under customary arrangements, while grazing land, forest and water are being managed through common property arrangements. Drought is recurrent, agricultural production is low and highly variable, and at least 600 000 people are considered 'highly vulnerable' to famine, denoting a condition in which modes of production and social and economic behaviour are being modified in response to food stress.

The communal lands tend to be located in drier areas least favoured for agriculture. Water shortage is the main problem reported. Thirty-two per cent of people still rely on relatively unsafe water from unprotected wells, rivers, streams and dams. There is a recorded density of only one protected well or borehole per 17 km^2. Of these, 40% fail during drought. For some the priority is for cleaner more reliable domestic water, for many it is for water to allow production.

In 1977 these areas were described as acutely over-populated with little or no land left for potential farmers; in some areas over 40% of men between the ages of 16 and 30 were landless. Now, decreasing access to arable land finds many families, particularly young, seeking alternative sources of livelihood that do not require large land holdings.

Farming systems are recognized to intensify in response to population increase. Irrigation is a common facet. The shift in use of groundwater from predominantly domestic use to production now being recorded is important and needs to be recognized in planning future rural water supply programmes.

communities would prefer, and how they might like to use the water to develop and protect their own livelihoods. The diverse range of small-scale production strategies that can be associated with a water point have not formally been promoted, and have actually been discouraged where considered to jeopardise the social objectives of domestic supply and sanitation. This is especially true in hard rock areas where conventional wells and boreholes are prone to failure and consequently have been developed as point sources for domestic water only. The misunderstanding of why wells and boreholes fail is discussed in Chapter 2.

Box 4: Informal small-scale irrigation

With the exception of the Nile Delta, large-scale irrigation was unknown in Africa until the twentieth century. In contrast, small-scale irrigation has been practised since time immemorial, in many varied forms according to local circumstances. The use of groundwater from hand-dug wells has been particularly common in semi-arid areas of Africa for irrigating small vegetable plots, especially by women, for whom it is a traditional activity.

Carter (1989) lists some of the problems that have beset large formal irrigation schemes in Africa. These include:

○ very high capital costs of large-scale land and water development;
○ inadequate financial resources for operation and maintenance;
○ lack of trained manpower and lack of irrigation experience among farmers on the schemes;
○ conflicts between the aims of government irrigation agencies and the aims of farmers;
○ overwatering leading to waterlogging and salinity;
○ low crop yields and poor returns both to the national economy and to the farmers;
○ difficulties of operating and maintaining mechanized crop production on smallholder schemes;
○ a range of environmental and health problems.

Underhill (1990) lists some of the comparative advantages of informal small-scale irrigation when compared to large irrigation schemes. These include:

○ implementation of informal irrigation can be considered as part of a development process rather than a one-off development action;
○ informal irrigation is based on self-reliance, schemes are farmer- or community-owned and managed;
○ informal irrigation rarely involves migration or resettlement; where possible, water resources are developed where rural people live;
○ relatively little infrastructure and low external inputs are required;
○ informal irrigation is highly adaptable and fits well into traditional farming systems that include rainfed crop production and livestock production;
○ as one of several agricultural activities, informal irrigation enables farmers to spread risk across a range of activities.

What are the benefits of productive water points?

The main benefits of productive water points, compared to conventional water points that provide domestic water only, are:

○   improved security of water supply

○ improved nutrition
○ income generation
○ diversification of livelihood strategies
○ positive feedback to the dryland farming system
○ community-based pump maintenance.

Improved security of water supply

Yields of boreholes and wells constructed through conventional water supply programmes are generally low and often unreliable, especially during drought. Table 1.1 shows typical yield figures recorded in an internationally funded drought relief programme in Zimbabwe in 1994. Almost half of the new boreholes drilled were dry or could not support a sustained handpump yield of $0.3 \, \mathrm{l \, s^{-1}}$.

Table 1.1 **Borehole siting success rates in Masvingo Province, Zimbabwe**
(minimum depth 45 m, average depth 64 m, maximum depth 121 m; *source: DDF Masvingo*)

District	Number of boreholes	Percentage of boreholes			
		Dry	<0.3 l/s	0.3-0.6 l/s	>0.6 l/s
Chivi	12	26	33	33	8
Zaka	11	8	46	46	0
Chiredzi	12	33	25	33	9
Bikita	12	25	50	8	17
Gutu	40	14	33	13	40
Masvingo	50	0	34	28	38
Mwenezi	12	17	25	33	25
Total	**149**	**13**	**34**	**25**	**28**

Table 1.2 **Distances (in metres) walked to productive water points (PWP) and other water sources**

Community	Members of PWP gardens		Non-members of PWP gardens	
	Ave. distance walked to other sources	Ave. distance walked to PWP	Ave. distance walked to other sources	Ave. distance walked to PWP
Muzondidya	733	825	438	
Gokota	388	1620	663	1500
Dekeza	588	780	600	643
Nemauka	436	887	443	850
Mawadze	300	1239	1208	3100
Matedze	336	979	267	1480
Mean	**463**	**1055**	**603**	**1515**

Box 5: Improved security of water supply

Productive water points provide improved security of water supply for a number of technical and social reasons.

Technical

○ Time is taken during water point siting to locate optimum aquifer properties. In some cases this will be achieved by exploratory drilling where present geophysical and traditional methods of siting have insufficient resolution in terrain of high spatial variability.

○ Well designs are chosen to suit the local hydrogeological conditions. In some cases, designs such as collector wells and large-diameter wells will be chosen to exploit the more consistent regolith aquifer rather than the potentially lucrative but highly unpredictable bedrock aquifer exploited by conventional boreholes.

○ Pump capacity is always matched to safe sustainable yield of the water point determined by pumping test and modelling of water table drawdown over an extended dry period.

○ Where well design allows, multiple handpumps are fitted to meet the required pump capacity and provide a back-up in the event that one pump breaks down.

○ User-friendly pump technologies are selected for which spare parts are locally manufactured, affordable and readily available.

Social

○ Community ownership of the water point and responsibility for operation and maintenance is promoted from the outset through participatory planning of management arrangements, a social contract, formulation with the community of a constitution and appropriate bye-laws, and an official handing-over ceremony.

○ Self-reliance is promoted through training in operation and maintenance and provision of basic tools and a simple pump-repair gantry at the time of scheme construction.

○ Production at the water point creates the incentive for community-based maintenance, while the income generated ensures ability to pay for repairs and buy spare parts when necessary.

○ Leadership is key, and emphasis is placed during scheme implementation on identifying natural leaders within the community willing and able to take the initiative forward.

In summary, productive water points are viewed as a long-term investment and they are purposely designed and implemented in a manner that ensures reliable water supply sufficient for both domestic use and production.

In contrast, figures recorded at other productive water points in the same area show more people using them in surges of usage of up to 54 per cent as other water sources fail during drought (Waughray et al., 1995). Their reliability is also emphasized by the fact that people are choosing to walk further to the productive water points (Table 1.2). The schemes are viewed as a time-saving intervention because people can rely on them to be working and to provide clean water relatively quickly. Despite being implemented at a time of perceived groundwater stress, these productive water points have continuously supplied yields of 10 000 to 20 000 litres of water per day, even during drought. The technical and social reasons for this improved security of water supply are summarized in Box 5 and detailed in the following chapters.

Improved nutrition

Surveys before and after construction of productive water points in Zimbabwe demonstrate the many advantages when schemes include irrigation of allotment-type gardens (see Box 6). One immediate benefit is improved nutrition. The gardens supply fresh vegetables that are consumed as a relish and sold to non-members of the scheme. Figure 1.2 shows how the scarcity of vegetables during normal dry seasons has been much reduced for both productive water point members and non-members. While some gardening continues during the rainy season (November to March) when rainfed cropping is taking place, garden production is highest, and returns greatest, during the dry season and during periods of drought.

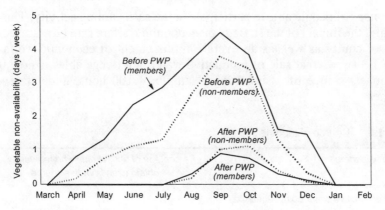

Figure 1.2 Vegetable non-availability before and after productive water points

Income generation

Productive water points possess considerable potential for income generation and income diversification for local families (Box 7). It is important to note not only the financial value of the income generated by the schemes, but also the *number* of people able to participate in the cash earning activity. This is

11

Box 6: Types of garden irrigation

In organizational terms, there are several different types of irrigated garden that can be found in dryland areas. These include: *private gardens* whereby one farmer or one household fence and manage their garden using water from their own well, from a river or a public water point; *collective gardens* whereby a garden is operated by the community as a co-operative and the produce is shared; and *community* or *allotment-type gardens* whereby a number of farmers or households have private plots within a communally-fenced area, water is obtained from a public water point, and produce belongs to the individual owners.

The type of gardening practised depends on a number of factors that include: availability of water resources, social and institutional factors, and advice given to communities by extension services. Availability of water is the major constraint on gardening in dryland areas, to such an extent that relatively few farmers are able to implement private wells and gardens either because there is insufficient groundwater beneath their land or because the cost of accessing groundwater is prohibitively expensive. Although there are some co-operative gardens, most rural communities, when given the choice, prefer some type of community or allotment-type garden where personal effort is rewarded directly.

the *number* of people able to participate in the cash earning activity. This is critical if the impact of the strategy on economic welfare is to be assessed in terms of equity as well as an overall improvement of economic welfare. Table 1.3 shows that sale of productive water point vegetables is now the commonest source of cash among a sample of 100 households at seven schemes.

Table 1.3   Source of income

Source of income	Percentage of households who regularly obtain cash from this source
Sale of productive water point vegetables	77
Off-farm remittances	59
Sale of rainfed crops	41
Beer brewing	36

Household surveys indicated that, of those who joined each community garden, 49 per cent were among the least wealthy in the community, and in 80 per cent of households women were the main decision-makers for the gardens. For women with little access to cash, materials or productive resources, obtaining a steady seasonal income from the schemes has greatly

Box 7: Some benefits of productive water points

After four years, surveys at seven schemes in Zimbabwe found that:

O 545 families were growing fresh vegetables, providing an average of 386 kg of produce per member household – equivalent to 58 tonnes/ha.

O Fresh vegetable consumption had increased, nutrition had improved, and the period of scarcity of fresh vegetables had been significantly reduced.

O 49% of garden members were among the poorest in the community, based on enumerator observation.

O 80% of the main decision-makers for the gardens were women.

O The average time spent by women in the gardens was 11 hours/week.

O The average internal rate of return (IRR) was 19% and average gross financial margins at farm gate prices was US$8400 per ha (assuming zero opportunity cost of dry season labour).

O 77% of garden members obtained a seasonal income as well as vegetables for home consumption.

O Average gross income from selling produce during 1996 was US$28 per member in an area where 50% of annual incomes were less than US$166.

O At least 911 households were also drawing domestic water from the seven schemes.

O This number could rise by up to 54% as other water sources failed each dry season.

O The sale of productive water point vegetables has become the commonest source of cash among a sample of 100 garden members.

O More than 70% of garden members use this new source of cash to create savings clubs and revolving funds which are being used to pay school fees and to buy inputs for other income-generating activities.

O Over 200 new household-based projects were reported to have started in part with cash generated by the productive water points, and ranging from livestock and fruit tree projects to pottery and knitting.

O The reliability rather than just the size of the income flows is said to be critical to the planning, scale and success of these new initiatives.

O There is positive feedback to the dryland farming system through creation of a reliable source of revenue used to buy rainfed farm inputs and engage labour. Importantly, it is younger, poorer members of the community who are keen to be hired, confident that cash is now available to pay for their work.

lowered elements of risk and income insecurity in the household decision-making and planning process (Box 8). This has been reflected in a sharp increase in both the perceived value of the plots and in their actual purchase price.

Another important aspect of the income-generating capability of productive water points is the extent to which they generate income flows that are reliable enough to allow the opportunity for savings and reinvestments in other non-

```
┌─────────────────────────────────────────────────────────────────────┐
│                      Box 8: Gender implications                       │
│                                                                       │
│  Role of women                                                        │
│  ○ main water collectors                                              │
│  ○ main gardeners                                                     │
│  ○ main users of vegetables                                           │
│  ○ main garden decision-makers within the family                      │
│                                                                       │
│  Positive impacts of productive water points                          │
│  ○ more vegetables                                                    │
│  ○ more income and financial independence                             │
│  ○ revolving funds                                                    │
│  ○ more reliable water (time saving)                                  │
│  ○ experience of community projects                                   │
│  ○ social interaction                                                 │
│  ○ increased role in community decision-making through the gardens    │
│                                                                       │
│  Negative impacts of productive water points                          │
│  ○ additional work                                                    │
│                                                                       │
└─────────────────────────────────────────────────────────────────────┘
```

farm activities. A good indication of this has been the revival and blossoming of 'revolving funds' at productive water points. A revolving fund, or *kukandirana* in Zimbabwe, is a traditional savings club operated and managed by a group of friends or colleagues. Each member of the group puts an equal amount into the fund at regular intervals, when the fund is then given to one member to spend provided the others in the group feel the purchases are justifiable. Each member takes a turn to spend the fund, hence the term 'revolving fund'. Where investment facilities are limited and where ready cash is important, membership of a revolving fund is a rational, risk-reducing exercise. It becomes possible only when members feel that they and their colleagues can rely on a steady source of income with which to participate.

Diversification of livelihood strategies

Surveys of seven productive water points indicated that in two to three years over 200 other household-based projects and numerous other group-based projects started up, at least in part due to the benefits accrued from the water points (Box 9). The household-based initiatives range from small-scale livestock projects and orchards to pottery, knitting and buying and selling second-hand clothes. Cash inputs to start these activities ranged from US$4 to US$44 per year. Other inputs provided by the productive water points included water and vegetable fodder. Annual income from these activities (gross financial margins at farm gate prices not including opportunity costs of labour) ranged from US$10 to US$262 per year.

14

Box 9: Creation of new projects

Productive water points in Zimbabwe have helped to start a large number of new agricultural and non-agricultural projects both by individual households and by groups. Factors noted by respondents to be particularly important to the success of the new projects were the *reliability* rather than just the size of the income flows generated by the productive water points, said to be critical to the planning, scale and success of the related group activities; and *access to information*, with lines of communication between external agencies and community groups being opened and maintained by village community workers, extension staff and local councillors. It was through this improved communication that many larger- scale group activities were initiated, with community money generated by the productive water points being matched or exceeded by other agencies.

New group projects started at productive water points

- 46 members of one scheme contributed US$2 each to buy a knapsack sprayer for use in the garden and on their rainfed cotton crops.
- 10 members at another site contributed US$2 each to make 10% of the total cost of purchasing a fence, chickens and feeding troughs. A GTZ programme contributed the remaining 90%.
- 46 of 50 members at one site created a crochet project, contributing US$230 per season (US$5 each) towards materials.
- 109 members at one site used productive water point water to mould bricks and garden income to build a house for a sewing-knitting project. The Kellogg Foundation provided other inputs.
- 15 members at one site established a poultry project for 75 broilers and a unit for 20 rabbits.
- Other community initiatives noted by respondents were the planting of fruit trees and woodlots, the establishment of seed nurseries, and several knitting, pottery and sewing clubs.

New projects initiated by non-members of productive water point gardens

- Buying vegetables at productive water point gardens and reselling at a profit at nearby townships.
- Payment by members for weeding, planting and harvesting in the productive water point gardens and in members' rainfed fields.
- Using water from the productive water point for beer brewing, pottery, mat-making and brick-making. Water-related production activities such as these were said to occur only where there was perceived to be 'spare' water remaining after the community's domestic and garden water requirements had been met.

Positive feedback to the dryland farming system

Positive changes in the local farming system occur as a result of productive water points in southern Africa, principally due to the creation of a reliable dry season income which is an important source of revenue to buy rainfed farm inputs.

○ Many members report that they use their garden income to buy more seed for their farms.

○ Garden members are engaging more hired labour inputs to plough, weed and spray their rainfed fields and to maintain contour bunds. Importantly, it is younger, poorer members of the community who are keen to be hired, and report that they engage more readily in labour contracts now as they are confident that the cash is available to pay for their work.

○ Garden income allows increased purchase of other farm inputs such as chemicals, fertilizers, hoes and spare parts for ploughs and cultivators.

○ Evidence of changes in cropping patterns include where private vegetable gardens have now been converted to fruit orchards. Vegetable supply is perceived to be met by the productive water point garden and fruit tree nurseries funded by the garden income are being set up.

Community-based pump maintenance

The greatest single threat to the success of rural water supply programmes is the question of operation and maintenance (O&M). Although water schemes are technically relatively easy to install, the short-, medium- and long-term problems of maintenance appear to be overwhelming. Government budgets are too small to cover the costs of maintenance, and in many countries water points lie broken because the local people lack sufficient capacity and incentive to restore a broken-down community pump. The proportion of broken-down installations is not well documented, and estimates vary from 10 per cent to more than 50 per cent. In Zimbabwe, for example, it estimated that about 8000 handpumps on communal water points are out of action at any one time.

In some countries, water policies now decree that the users themselves should be responsible for maintenance, but this relatively new concept of self-sufficiency has so far achieved limited success. Village-level operation and maintenance (VLOM) costs can be high, and ways of collecting sufficient revenue at conventional domestic water points have still to be worked out.

However, when a water point is given an economic value through production it creates both the incentive and the income needed to make community-based pump maintenance a reality. At productive water points in Zimbabwe, the pumps have either been maintained locally from the outset or the users have chosen to pay government pump-minders to carry out repairs. At all

16

schemes, the users themselves are choosing to buy spare parts (and pay pump-minders) using income generated by the schemes; they no longer rely on external assistance. Although the payment regimes at the productive water points have been somewhat arbitrary (Box 10), ranging from US$50–245 per year, they are not insignificant and are in stark contrast to conventional rural water supply programmes where the annual cost to the state (or donor) is of the order of US$90, which equates to an additional (discounted) cost of US$1120 per domestic water point over a standard lifetime of 20 years.

What problems can emerge at productive water points?

Experience suggests that productive water points can improve nutrition, health and income and provide a stream of other social, financial and local institutional benefits. However, some special management problems do arise when a common property resource such as water assumes an economic value through production, and committees at productive water points can face some difficult management decisions. A number of issues is emerging which can impact on the institutional sustainability and performance of productive water points.

Local property rights to land, water and other natural resources

The question of who owns a common property resource such as land, water, woodland or wildlife is compounded when the resource assumes an economic value and becomes a source of income. The problem is particularly acute in community-based projects that require land – and the allocation of land for productive water points, although relatively small areas, is no exception. Conflict can arise, for example, where an influential person may offer land (and the groundwater beneath) for a productive water point but later undermines the role of the water point committee by refusing right of access to some user groups, or where a considerate person offers land (and the groundwater beneath) for a productive water point on the promise of compensation, but this promise is subsequently broken. As we shall see, institutional support and arbitration from higher levels of authority are important.

Equity among different user groups

The users of productive water points fall into three broad groups: households involved in scheme construction who generally use the water point as a source of water for domestic use *and* income-generating activities; households probably not involved in construction who use the water point primarily as a source of domestic water; and a more peripheral group that use the water point for domestic supply in times of drought when other water points fail. Each group can apply pressure for their own production systems enhancement, particularly as the demand to expand and initiate new projects increases with time.

The different user groups operate on different spatial and temporal scales. The first group, with the productive activity as their common interest, are able to lobby effectively through their productive water point committee which generally includes key members of the community who have experience in village-level administration. The group focuses on the productive activity and its effect on their farming and livelihood strategies, thinking ahead as income security increases. In contrast, the second and third groups focus more on the immediate supply of water, having less incentive to consider or contribute to long-term management issues at the productive water point. Their viewpoints are expressed either by prominent individuals or by non-participation in activities such as pump repair. The third group have even weaker lobbying power and may play little part in management decisions at the productive water point. Depending on location, a fourth user group may include schools, businesses or individual entrepreneurs, and such groups can exert significant lobbying power on the use of the productive water point. As we shall see, enabling each of these user groups to meet and discuss their access to the opportunities for production is critical.

Planning, management and membership of new projects

As new projects are proposed at productive water points, new committees may be formed. To avoid conflict, new committees should clarify their relationship with the existing productive water point committees, perhaps with mediation by traditional leaders and local authorities. However, if new initiatives come from non-members of the original project, the original members may be reluctant to see any change to their investment. In cases such as these, trained facilitators and agricultural extension staff can play an important role in providing the necessary information to help the productive water point committee make difficult management decisions and gain in the confidence needed for local resource management.

What are the criteria for a successful productive water point?

To be successful, a productive water point should:

- be reliable and supply at least 10 000 litres of water every day, especially during drought;
- serve an immediate community that has a high level of dependence on the resource;
- generate an income stream not necessarily large but which is reliable and available to all participants;
- be owned, managed and maintained by the community rather than the state.

The community, through their productive water point committee, should:

- formulate and agree on a constitution for the development and management of their productive water point, which promotes social cohesion by providing guidance on the contentious issues of equity, membership, participation of the poor, user fees, cost-recovery for operation and maintenance, reinvestment of revenue, and rights of access particularly for non-members of the initial scheme wishing to use water and land to start new income-generating activities;
- be clear on the roles and responsibilities of local authorities and support agencies in providing institutional support and arbitration from higher levels of authority, guided by appropriate bye-laws where necessary.

The community, through their local extension service or NGO, will need access to relevant technical information to help optimize productive potential, on issues that will include:

- sustainable yield of the water point and corresponding irrigable area;
- appropriate pump technology options;
- pest and disease control;
- cropping patterns which take account of market demands;
- irrigation methods and irrigation scheduling;
- other productive activities such as livestock and dairy projects;
- protection of the groundwater resource.

The following chapters discuss the steps taken during programme development with the community to help ensure that these criteria are met and that the necessary information is provided, starting with the all-important question of how to secure continuous supplies of groundwater sufficient to develop productive water points in drought-prone areas.

References and further reading

Anon. (1992). *President's report on the Drought of 1991-92*. Min. Public Information, Harare.

Boserup, E. (1965). *The Conditions of Agricultural Growth: The economics of agrarian change under population pressure*. Earthscan Publications, London.

Campbell, B., du Toit, R. and Attwell, C. (1988). *Relationships between the environment and basic needs and satisfaction in the Save Catchment.* Univ. of Zimbabwe, Harare, 119pp.

Carter, R.C. (1989). *NGO Casebook on Small Scale Irrigation in Africa*. AGL/MISC/15/89, FAO, Rome.

Cleaver, K. and Schreiber, G. (1994). *Reversing the Spiral: The population, agriculture and environment nexus in sub-Saharan Africa.* World Bank, Washington DC.

De Lange, M. and Crosby, C. (1995). Towards successful small-farmer irrigation. *South African Water Bulletin*. September.

Ellis, F. (1998). Livelihood diversification and sustainable rural livelihoods. In: *Sustainable Rural Livelihoods – What contribution can we make?* Ed. D. Carney. Department for International Development, London. 53-66.

IPTRID (1999). Poverty reduction and irrigated agriculture. *Issues Paper No. 1*, FAO, Rome. 21 pp.

Magadza, C.H.D. (1992). Water resources conservation. *Zimbabwe Science News* 26(10), 82-87.

Moriarty, P.B. and Lovell, C.J. (1998) Groundwater resource development in the context of farming systems intensification and changing rainfall regimes: a case study from south-east Zimbabwe. *AgREN Network Paper 81*, 15-21, ODI, London.

Riddell, R. (1978) The land problem in Rhodesia. *Socio-Economic Series No. 11*, Mambo Press, Gweru, Zimbabwe.

Scoones, I. (1996). *Hazards and Opportunities: Farming livelihoods in dryland Africa – Lessons from Zimbabwe.* Zed Books, London.

Smout, I. (1990). Farmer participation in planning, implementation and operation of small-scale irrigation projects. *ODI/IIMI Irrigation Management Network Paper 90/2b.* Regent's College, London

Uhl, V. W. and Atobrah, K. (1987) A global view of the hydrogeology of crystalline rocks. *World Water Conference, Nairobi vol. 1*, 1-25.

Underhill, H.W. (1990). *Small-scale Irrigation in Africa – in the context of rural development.* Cranfield Press, Bedford, UK. 90pp.

USAID (1994). *An Assessment of Vulnerability in Zimbabwe's Communal Lands.* Famine Early Warning System Project (698-0466), US Agency for International Development, Harare 42pp.

Waughray, D.K., Mazhangara, E.M., Mtetwa, G., Mtetwa, M., dube, T., Lovell, C.J. and Batchelor, C.H. (1995). Small-scale irrigation using collector wells: Pilot project – Zimbabwe. Return-to-households survey. IH Report ODA 95/13, Institute of Hydrology, Wallingford, UK. 107pp.

2   Understanding why wells and boreholes fail

Underground water explained

THIS SECTION IS MEANT as a guide for the layman. It provides some basic information on the factors that determine groundwater availability and the reasons why conventional wells and boreholes fail.

Hydrology

Water exists in the ground at some depth nearly everywhere on Earth. Almost all of the world's stock of fresh water, 8.2 million cubic metres, or more than 97 per cent of the total available, is inside the Earth itself and exists as groundwater. The rest exists in lakes, streams and rivers.

Figure 2.1 illustrates the hydrological cycle in a dryland catchment. Gravity pulls water from the skies to beneath the ground surface, distributes it among permeable layers of the Earth and influences the direction in which it flows. Water first enters a zone of aeration of soil and decomposed rock containing both water and air. The depth of this unsaturated zone varies widely, from a few centimetres where the water table is high to hundreds of metres in areas with a low water table. Some water sinks to the saturated layers beneath, some is absorbed by plants and transpired, and some evaporates from the soil surface into the air again. The lower zone of saturation is the Earth's main reservoir. Wells and boreholes tap into it and springs, rivers and lakes are its natural outcroppings on the surface. The sparkle of water at the bottom of a well is the top of the saturated zone or water table. Around it – and continuous with it — the water table extends to the extent of the aquifer.

Most underground water is constantly in motion, being pulled by gravity from higher aquifers to lower. It is lifted from the ground by plants and by man. Without human impact, recharge to the groundwater 'reservoir' is balanced by natural discharge or recession through spring flow, base-flow to streams, deep percolation into underlying fractures, evaporation from seepage zones and lakes where the water table reaches the surface, and transpiration from areas where the water table lies within the rooting zone of vegetation. Exploitation of groundwater by means of boreholes and wells means that some groundwater is being diverted from being discharged naturally.

In principle, a sustainable level of abstraction from an aquifer is one that does not exceed the recharge to groundwater. In practice, abstraction should

21

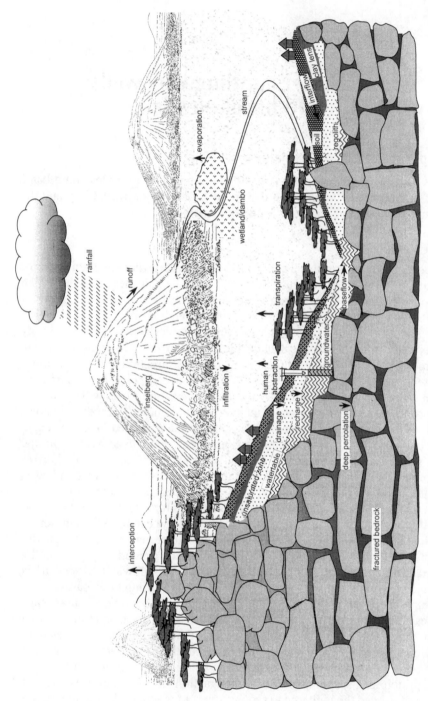

Figure 2.1 The hydrological cycle in a dryland catchment on crystalline basement rock

be less than recharge because of the relationships between available groundwater and other components of the hydrological cycle. Over-exploitation of groundwater is dealt with in Chapter 7.

Geology

The geology or origin, structure and composition of the Earth's surface is the key to groundwater availability. It influences both the soil type and surface features of the particular landscape (which in turn affect the hydrology) and the weathering profile and nature of the openings in the rocks which hold and convey any groundwater.

There are three major rock groups:

○ Igneous rocks that have formed from a molten state. Some types solidified at great depths below the surface are referred to as intrusive, examples of which are granite, gabbro and dolerite. Others formed from lava or volcanic ash ejected on the surface, examples of which are basalt, rhyolite and tuff.

○ Sedimentary rocks that have formed by deposition of sediment from water, ice or air. Through time, deep burial, compression and chemical changes, they have become consolidated. Examples are sandstone, shale, mudstone, dolomite and limestone.

○ Metamorphic rocks formed from both igneous and sedimentary rocks when subjected to great heat and pressure below the earth's surface. Granulite, (meta-)quartzite and various gneisses are examples.

The Earth's crust is dynamic and changing, albeit very slowly over millions of years. While in certain parts of the world, plateaux and mountains were being made and worn down, in other parts the land became submerged beneath the sea for long ages and continued to sink slowly while mud and sand derived from eroding land built up deposits thousands of metres thick. These loose materials were converted by compaction into rock, eventually to be uplifted above the sea to form either elevated plateaux or mountain ranges, the beginning of another cycle of erosion and deposition, of sedimentary rock formation, igneous activity and uplifting, a process repeated many times during the Earth's history. Consequently, groundwater conditions vary, both on a regional and local scale, depending on the basement geology, tectonic history of the locality, climate past and present, and local relief.

Groundwater

In unconsolidated sedimentary deposits such as alluvial sand and clay, the openings in which groundwater occurs are the pores between the mineral grains. These are termed primary openings. Upon burial these interstices are reduced in size, and are completely destroyed in older sedimentary rocks which have been deeply buried and have undergone some degree of metamorphism. Similarly, intrusive igneous and metamorphic rocks, having been formed at depth, do not contain any appreciable primary openings. These compact rocks and aquifers, lacking primary porosity, are termed 'hard rock aquifers'.

All rock types, during the course of geological time, are subjected to vertical and horizontal stresses, resulting in the opening of sets of horizontal, dipping and vertical fractures known as joints, faults and cleavage planes. A joint is a parting plane which separates two parts of a once continuous block and along which there has been no visible movement parallel to that plane. Faults are ruptures along which the opposite walls have moved past each other. Faults may be accompanied by breccia, which consist of angular fragments of crushed rock. If not cemented subsequent to faulting, these breccia may act as aquifers. The near-surface opening of these fractures allows the ingress of air and moisture. Weathering commences, resulting in disintegration and decomposition of the rocks.

The fractured rock, termed bedrock, may be exposed on the surface in places. Elsewhere it is covered by a mantle of soil, clay, sand or gravel transported from elsewhere, or by material derived through in-situ weathering of the underlying bedrock (Fig. 2.1). The thickness of this weathered layer or regolith is highly variable. Importantly, it contains fractures and pores, termed secondary openings as they were created after the rock formation. It is these regolith fractures and pores that provide space for the storage and transmission of groundwater in what would otherwise be compact 'hard rock'.

Groundwater in transit through any aquifer may be viewed as groundwater in storage. The volume of groundwater in storage at any time determines the aquifer potential, and should be sufficient for a supply rate equal to the long-term mean recharge rate to be maintained until the next recharge event, and if necessary beyond that event, should recharge on that occasion fall short of the mean. In hard rock formations, the volume of water in storage, and hence the aquifer potential, is limited. It depends on the extent to which the zone of secondary porosity has been created by weathering processes, and then primarily on water stored in the secondary intergranular openings in the decomposed rock rather than in fractures, the storage coefficient of decomposed rock being at least one order of magnitude larger than that of fractured rock, regardless of fracture density.

In hard rock areas, the aquifer therefore depends on the depth of saturated weathering, and this is highly variable. Some parent rock minerals are more susceptible to decomposition than others, so have different weathering properties. Structural features such as joints, faults and dykes also have an important effect on weathering. Other factors affecting the thickness of the regolith are climate (weathering is more intense under wet humid conditions) and concentration of runoff in which steepness of surface features such as rock outcrops play an important role.

If water is to move through an aquifer, the pores and fractures must be connected to one another. The property of rocks to transmit water is known as the permeability. The larger the interstices through which the water flows, be they pores or fractures, the higher the permeability of that rock. For instance, clay is highly porous and stores high volumes of water but its permeability is very low because of the fineness of the clay particles. A coarse sand, on the other hand, has lower porosity but its permeability is high. The permeability of the regolith, composed predominantly of clay minerals, is very low. Movement of groundwater in hard rock formations depends primarily on the lateral and vertical extent of the fractures, and the connectivity of these fractures to the secondary porosity of the regolith. High-yielding wells and boreholes are those that pass through a significant depth of saturated regolith of reasonable permeability, or which are well connected to this resource via an extensive network of fractures.

Potential of the groundwater resource

Approximately two-thirds of the world's dryland areas are underlain by hard rock aquifers, principally in crystalline basement rocks of igneous or metamorphic origin. Figure 1.1 shows their distribution in Africa, Asia and South America. It is important to note that these aquifers have low groundwater potential relative to other aquifer types because of the low volume of water stored in hard rock formations. However, in many dryland areas the water stored within these rocks is, for many people, the *only* source of water for long periods of time. Thus, hard rock aquifers are vitally important.

To understand how productive water points can safely be developed on hardrock aquifers, we must look at the current utilization of groundwater as a proportion of recharge, and then at the reasons why conventional wells and boreholes often fail at present.

Current utilization of groundwater as a proportion of recharge

In most dryland areas groundwater is considered to be an underutilized resource, with future dangers derived more from declining quality of the water than from the quantity. Table 2.1 shows the degree of utilization of groundwater in various countries. Percentage usage is typically domestic (8), industry (23), and agriculture (69). On an international basis, major users are (in $km^3 \, y^{-1}$): India 150; USA 101; China 74; USSR 45; Pakistan 45. In other countries, all types of use are typically less than 15 per cent of the total available.

Groundwater use by rural populations in dryland areas is far less (Box 11). Regional studies in Africa, for example, show that although variable in space and time, current use of groundwater is typically less than 4 per cent of long-term mean annual recharge and mean annual recharge is of the order 2–9 per cent of mean annual rainfall. Figure 2.2 shows regional estimates of groundwater recharge across South Africa, a predominantly dryland country underlain by various hard rock aquifers.

Table 2.1 Groundwater abstractions in countries of Africa, Asia and South America (*source:* Margat, 1991)

Country	Year	Withdrawals $(10^9\ m^3\ y^1)$	(%)	Country	Year	Withdrawals $(10^9\ m^3\ y^1)$	(%)
Africa				**Asia**			
Egypt	1985	3.4	6.0	Bangladesh	1979	3.4	-
Madagascar	1984	4.76	29.0	China	1985	74.6	15.0
Mali	1989	0.1	-	South Korea	1985	1.2	8.0
Morocco	1985	3.0	27.0	India	1979	150.0	35.0
Mauritania	1985	0.88	47.0	Iraq	1985	0.2	0.5
Niger	1988	0.13	9.0	Israel	1986	1.2	67.0
Senegal	1985	0.25	-	Pakistan	1980	45.0	29.0
Somalia	1985	0.3	3.5	Syria	1985	3.7	14.0
South Africa	1980	1.79	13.0	South Yemen	1975	0.35	18.0
Sudan	1985	0.28	1.3	**South America**			
				Argentina	1975	4.2	15.0
				Peru	1973	2.0	-

Box 11: Estimating current groundwater use: an example

In 1993 there were 2161 boreholes and 1225 deep wells recorded in Masvingo Province, Zimbabwe, of which 756 and 610 respectively were known to fail during drought. Each water point is, or will be fitted with, a single handpump. Assuming that all 3386 water points yield water and the average yield of a handpump is 6500 m³ y⁻¹ (a pumping rate of 0.5 l s⁻¹ for 10 hours per day for 360 days per year) the maximum groundwater abstraction for the province will be 22×10^6 m³ y⁻¹. A conservative estimate of recharge for the drier parts of Masvingo Province is 2 per cent of rainfall (Houston, 1988). Assuming average annual rainfall of 500 mm, recharge of 10 mm y⁻¹ over the province may be expected. The area of the province is 57 000 km², giving an annual average recharge of 570×10^6 m³ y⁻¹. The current use of groundwater in Masvingo Province is thus of the order of 4% of annual average recharge.

Mean annual rainfall and regional estimates of recharge are rather misleading terms in dryland areas, given the high inter-annual and spatial variation. In years of above average rainfall, recharge can be considerably higher than the long-term mean and the fraction of groundwater used is even less. Table 2.2 shows the annual water balance measured for a small (4.6 km²) instrumented catchment in a communally managed dryland area of Zimbabwe with mean annual rainfall 700 mm. Figure 2.3 shows monitoring data recorded at the productive water point in this catchment. Recharge rates up to 300 mm per year have been measured in wet years. However, natural groundwater recession, principally through abstraction by deep-rooted

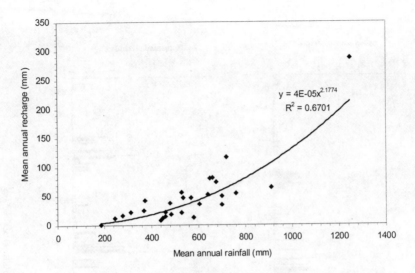

Fig. 2.2   Regional estimates of groundwater recharge across South Africa
(*source of data:* Vegter, 1995)

Table 2.2   Annual water balance of Romwe catchment, Zimbabwe (year 1 July –
30 June, units in mm) (*source*: Lovell *et al.*, 1998)

	Runoff	Recharge	Change in groundwater storage	Natural groundwater recession	Human use	Balance to evaporation, change in soil moisture storage and other losses
1994/95 rainfall 738 mm	4	38	-34	72	1	695
1995/96 rainfall 990 mm	93	262	+100	162	1	634
1996/97 rainfall 1140 mm	84	296	+62	234	1	759

vegetation, lateral flow and leakage into the fractured bedrock, can directly account for up to 230 mm of this recharge, while human abstraction for domestic use and irrigation by 103 families living in the catchment is trivial at less than 1 mm per year of recharge. In this environment, groundwater is not failing because of resource use and the number of water points could safely be increased tenfold and still have negligible impact on the natural recession of groundwater.

The difficulty of developing basement aquifers has resulted in the use of wells and boreholes as point sources for domestic supply, fitted with a handpump, bucket pump or bucket and windlass. In 1987, the Commonwealth Science Council reported that total abstraction for a rural community with

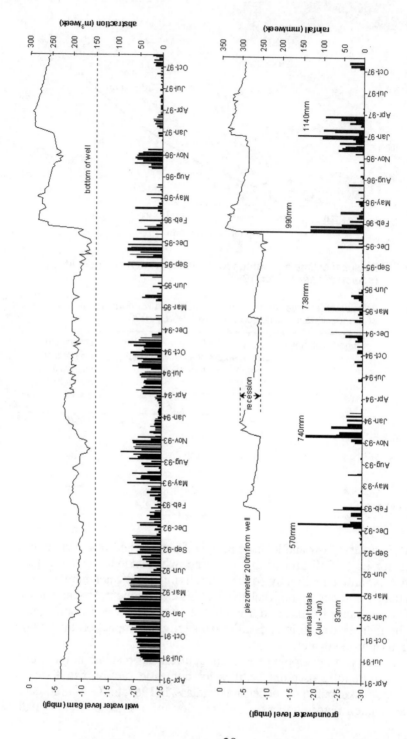

Figure 2.3 Water levels, abstraction and rainfall recorded at Romwe productive water point, 1991–97

28

this degree of development will amount to the equivalent of 0.5 mm of recharge, where present evidence suggests that actual recharge is commonly much higher, in the range 50–150 mm. They recommended that at least some of this surplus could safely be utilized for higher demands, if development methods could be devised which are feasible and economic. Although recharge is important for stream flow and ecology, and examples of non-sustainable levels of abstraction do now exist (for example, in India where borehole finance schemes and rural electrification have led to overdevelopment) groundwater in many dryland areas still offers potential for greater use if the aquifers can be appropriately developed and managed.

Why do wells and boreholes fail?

Conventional wells and boreholes fail due to a number of factors, acting alone or in combination. These factors, and the countermeasures that can be taken, are illustrated in Figure 2.4 and summarized in Table 2.3. Generally, it is technical difficulties and social constraints rather than overall groundwater

Table 2.3   The principal causes of water point failure in hard rock aquifers and the countermeasures that can be taken

Cause of failure	Countermeasure
REGIONAL • high natural groundwater recession a feature of basement aquifers and severe in extended dry cycles when recession exceeds recharge	• improved siting to locate the water point in and connected to the greatest depth of saturated weathering
LOCAL • drawdown of the water table immediately around the well or borehole on pumping due to low aquifer permeability and/or poor connectivity to water stored in the regolith	• improved siting to identify good permeability/connectivity to water stored in the regolith • selection of well designs suited to low permeability conditions • manage abstraction to suit individual water point performance
SOCIAL/INSTITUTIONAL • inadequate arrangements for pump maintenance and repair	• empower the local community to own the resource and assume responsibility for O&M • give the water point an economic value through production
• drawdown of the water table immediately around the well or borehole due to interference by other wells and boreholes in the area	• promote collective action at fewer well-sited public water points with planning and legislation to prevent ad hoc development of multiple private wells

29

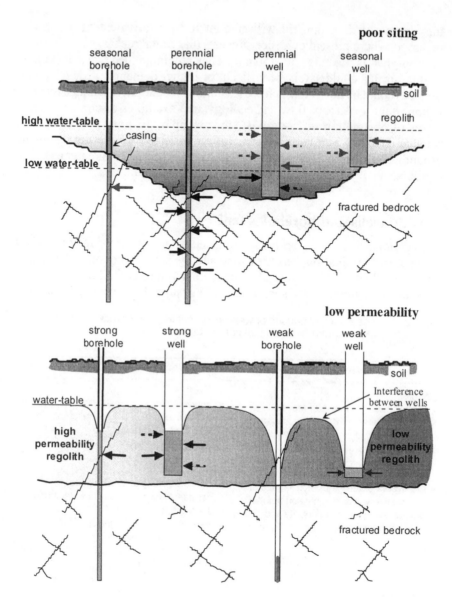

Figure 2.4 Well and borehole failure mechanisms. With poor siting comes poor connectivity between regolith and bedrock and only shallow depths of saturated weathering. These cause low yields and failure as water tables fall annually through natural recession and even more severely during drought. The problems are compounded in low permeability conditions or where abstraction rates are too high. Localized water table drawdown on pumping restricts water entering from the surrounding aquifer, and is manifest as a general decline where too many water points are sited close together and interfere with one another.

availability which cause conventional water points to fail, and which explain why communities remain short of water and groundwater use remains low.

Low rainfall / high recession

Conventional wells and boreholes fail if they are inadequately sited or of inappropriate design to cope with the low rainfall, high natural groundwater recession, high spatial variability and low permeability that characterize dryland areas and hard rock aquifers. Figure 2.5 shows the annual and cumulative departure from long-term mean annual rainfall recorded in dryland areas of Zimbabwe and Zambia. Although there is no evidence of an overall decline in rainfall, the trends indicate cycles of above and below average

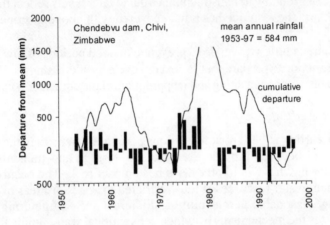

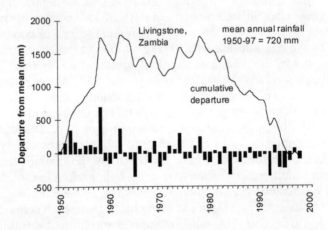

Figure 2.5 Departure from mean annual rainfall in dryland areas of Zambia and Zimbabwe

rainfall. When rainfall is generally low for an extended period (for example, 16 years in Zambia from 1979 to 1995 and 13 years in Zimbabwe from 1980 to 1993) groundwater recharge fails to match natural groundwater recession (which remains relatively constant from year to year) and groundwater levels fall. This fall in water level leads to many traditional sources drying up, since these tend to be sited convenient to the homestead rather than in optimum depths of saturated weathering, and people are forced to walk long distances to more reliable water points.

Modelling catchment hydrology in Zimbabwe using rainfall records from 1953 to 1999 clearly shows that the reason for low groundwater levels and water point failure in the early 1990s was the extended period of low rainfall from 1980 to 1993 and not human impact on hydrology through either land-use change or groundwater abstraction. Groundwater levels reflect the cycles of rainfall, and the resource has now recovered with above average rainfall since 1995.

While there is little we can do to prevent extended periods of low rainfall, we can take action to prepare for the next dry cycle by addressing the problems of inadequate water point siting and inappropriate choice of water point design.

Inadequate siting

As we saw earlier, in hard rock formations pores and fractures in the regolith and fractured bedrock provide the space for storage and transmission of groundwater in what would otherwise be compact rock. The occurrence of viable groundwater resources is thus restricted to zones or areas of deepest weathering, reasonable permeability and interconnected fracturing. Figure 2.4 illustrates the mechanisms by which conventional water points fail when these conditions are not satisfied. They fail due to natural groundwater recession during extended dry periods if sited in shallow depths of saturated regolith. They also fail if sited in aquifers of low permeability due to drawdown in the water table immediately around the well or borehole upon pumping. To avoid failure through natural groundwater recession, water points should be sited in the maximum depth of saturated weathering. To avoid failure through localized drawdown of the water table, water points should be sited in aquifers of open fracturing and reasonable permeability.

That some reliable water points remain in most areas even at the height of an extended dry cycle is evidence of the importance of careful water point siting. In fact, improved siting is the key to reliable water supply in hard rock aquifers. Unfortunately, Figure 2.6 illustrates the spatial variability in groundwater conditions that characterizes crystalline basement aquifers. This variability was recorded in a transect of only 700 m across Romwe catchment in Zimbabwe, an area of younger, undifferentiated gneisses. It helps to explain why it is so difficult to site reliable water points in this terrain. This is especially true where conventional wells and boreholes are sited with respect to the homestead rather than to the groundwater conditions. The average

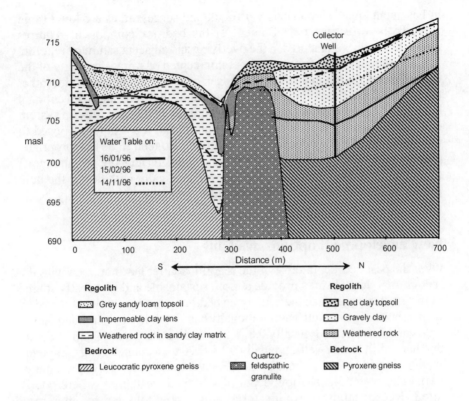

Figure 2.6 The spatial variability in groundwater conditions recorded in a transect across Romwe catchment, Zimbabwe
(redrawn from Moriarty and Lovell, 1998)

abstraction from family wells in Romwe is only 210 litres per day and yet many still fail each dry season due to natural groundwater recession and localized drawdown of the water table.

In this environment, test drilling and simple pumping tests have been found helpful to locate optimum groundwater conditions for water points because neither traditional water divining nor present geophysical techniques have sufficient resolution in terrain of such high spatial variability. This alternative approach to water point siting is discussed further in Chapter 5.

Inappropriate choice of well design

Hard rock aquifers may be classed as two-layer systems, the upper weathered layer or regolith which stores water and the underlying fractured bedrock which stores little water. In conventional water supply, the fractured bedrock is exploited either by deep boreholes typically 60–80 m in depth, 150 mm in diameter and cased in the regolith, or by deep wells extended beyond the weathered zone often using dynamite. Siting to intercept bedrock fractures in basement areas is notoriously difficult and success rates are low (Table 1.1). To be effective, development of the bedrock component requires interaction with water stored in the overlying and adjacent saturated regolith. Where interconnected fissures are not intercepted or storage capacity of the immediate regolith is inadequate, well deepening will not be successful (other than to increase the volume of water stored in the well) and conventional borehole yields will be low and unreliable. In the two-layer system characteristic of hard rock aquifers, casing in the regolith (as opposed to screening) and drilling deeper than 40–50 m also tend to be inappropriate. Casing denies access to water from the regolith and the frequency of intercepting bedrock fractures decreases dramatically with depth, the bulk of the readily accessible and exploitable groundwater generally being contained within the first 40 metres.

New development options available

Over the past decade, interest in the regolith aquifer has increased with the recognition that this may provide a more sustainable and less costly source of rural water supply than the underlying bedrock fractures previously targeted by boreholes. The regolith has traditionally been exploited by hand-dug family wells and deep wells, typically 0.8–1.2 m in diameter and 5–20 m in depth, and hand-drilled tubewells, typically 0.15–0.20 m in diameter and completed in the regolith; yields of both are generally low, limited by low aquifer permeability and the shallow depth of saturated weathering where poorly sited. Recent studies provide three new methodologies for increased abstraction of water from the regolith:

- screened regolith boreholes sited by test drilling and completed to 40 m
- large-diameter wells sited by test drilling and dug to bedrock
- large-diameter collector wells sited by test drilling, dug to bedrock, and completed with horizontal boreholes drilled at the base of the well.

Figure 2.7 illustrates these designs, together with conventional deep boreholes, deep wells, family wells and hand-drilled tubewells. Table 2.4 shows average yields recorded for each design in various projects. Essentially, the new designs are sited by test drilling to locate the maximum depth of saturated weathering of reasonable permeability, and are constructed to make effective use of the regolith aquifer by screening (rather than casing) and by large diameters (2-3 m) which, in low permeability conditions, provide sufficient storage for water as the wells fill slowly each night. The horizontal boreholes of the collector well (or Rannay-type well), drilled radially from the base of a large-diameter well in several directions to a distance of up to 30 m, are designed specifically to overcome high spatial variability and low permeability where these are particular constraints.

Due to non-rigorous testing of family wells and deep wells in past programmes, and the limited sample size of the new options, it is too early to draw definitive conclusions, but some important lessons are emerging:

○ Development of the fractured bedrock by boreholes requires interaction with water stored in the overlying and adjacent regolith. If interconnected fractures are not intercepted, low yields result; about half of all new boreholes drilled are dry or yield less than 5–10 m^3 d^{-1}. However, where interconnected fissures are intercepted, some very high yields >150 m^3 d^{-1} may be obtained. Consequently, some existing boreholes are grossly underutilized where single handpump capacity is far below the safe yield of the water point.

○ Siting to intercept bedrock fractures in basement areas is difficult and success rates are low. Siting boreholes by shallow exploratory drilling in the regolith to locate greatest depths of saturated weathering appears to help overcome this variability found at depth.

○ Boreholes in hard rock aquifers would be better if screened in the regolith rather than cased.

○ Average yields of screened regolith boreholes, large-diameter wells and collector wells sited by test drilling, although lower than conventional deep borehole yields at some locations, are far more consistent and quite adequate, providing upwards of the 10–20 m^3 d^{-1} of water needed to support domestic use and small-scale irrigation.

○ Although constructed at the peak of perceived groundwater stress in the early 1990s, collector wells sited by test drilling to locate maximum depths of saturated weathering and completed with horizontal boreholes to overcome low permeability, have sustained yields greater than the sum of abstraction from all family wells and have supported small-scale irrigation for up to 130 families at each site.

The relative importance of the regolith and underlying fractured bedrock, and the most appropriate development strategies, will vary from region to region (Box 12). When planning productive water points it is important, therefore, to consider all potential sources of water including existing under utilized resources, and if siting new water points, to select the most appropriate designs suited to the local groundwater conditions. Surveys undertaken in drought-prone areas of Zambia and Zimbabwe, where development agencies are presently sinking many conventional boreholes and wells largely irrespective of local groundwater conditions, indicate that nearly half of all existing water points are either under-utilized or of inappropriate design (Box 13).

Figure 2.8 summarizes ways to increase groundwater availability. At least six of these can be used in hard rock terrain. Surface water resources can also be developed for production. The challenge is to select the most appropriate technology (Box 14). Cost is one factor. The next chapter compares costs and benefits of the various development options now available.

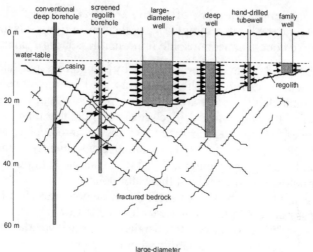

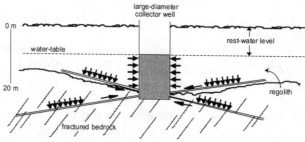

Figure 2.7 Seven water point designs used for groundwater abstraction in hard rock aquifers

Table 2.4 Average yields ($m^3 d^{-1}$) of different water point designs

Design	Sample size	Average yield	Range	Standard deviation
Family well[1], average depth 10.8 m	110	1.9	0.0 – 7.4	1.7
Deep well[1] average depth 17.6 m	110	7.1	0.1 – 16.5	3.7
Conventional deep borehole, avg. depth 64m, cased in regolith[2]	149	22.5	0 – 162.0	30.2
Screened regolith borehole to 40 m, sited by test drilling[3]	5	36.4	26.0 – 45.0	7.1
Large-diameter well to bedrock, sited by test drilling[3]	9	25.1	10.2 – 55.3	15.1
Collector well to bedrock, sited by test drilling[3]	8	29.5	10.8 – 62.5	15.9

1 DFID and WaterAid, Bikita District, 1994 Emergency Water Supply Programme. No pumping tests. Bucket counts

2 World Bank, Masvingo Province, 1994 Drought Relief Programme. 5-hour pumping tests. 10-hour pumping day assumed.

3 Lovell et al. (1996) 5-hour pumping tests and modelling to project sustainable yield during drought .

36

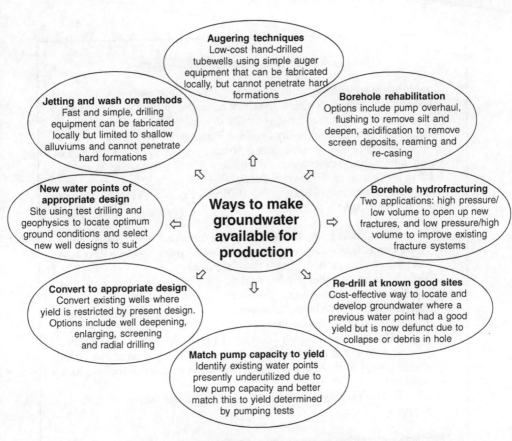

Augering techniques
Low-cost hand-drilled tubewells using simple auger equipment that can be fabricated locally, but cannot penetrate hard formations

Jetting and wash ore methods
Fast and simple, drilling equipment can be fabricated locally but limited to shallow alluviums and cannot penetrate hard formations

Borehole rehabilitation
Options include pump overhaul, flushing to remove silt and deepen, acidification to remove screen deposits, reaming and re-casing

New water points of appropriate design
Site using test drilling and geophysics to locate optimum ground conditions and select new well designs to suit

Ways to make groundwater available for production

Borehole hydrofracturing
Two applications: high pressure/low volume to open up new fractures, and low pressure/high volume to improve existing fracture systems

Convert to appropriate design
Convert existing wells where yield is restricted by present design. Options include well deepening, enlarging, screening and radial drilling

Re-drill at known good sites
Cost-effective way to locate and develop groundwater where a previous water point had a good yield but is now defunct due to collapse or debris in hole

Match pump capacity to yield
Identify existing water points presently underutilized due to low pump capacity and better match this to yield determined by pumping tests

Figure 2.8 Eight ways to increase groundwater availability for production

---

Box 12: The importance of regolith and bedrock

The importance of the regolith and the underlying fractured bedrock components of hard rock aquifers depends on the basement geology, tectonic history of the locality, climate, and local relief. Bedrock in Malawi rarely forms a significant aquifer while, in Uganda, it seems that it is the regolith which stores and transmits most groundwater and thus be the key to future development. In Botswana, the weathered zone only occasionally constitutes an aquifer worth developing, but good yields were obtained from fissured zones in the bedrock. In Zimbabwe, both regolith and fissured bedrock aquifers occur. Viable regolith aquifers are found on younger undifferentiated gneisses and intrusive granites; on older gneiss complexes the depth of saturated weathering is generally too small and/or the depth to water table too great for large-diameter or collector wells. On Beitbridge paragneiss, potential regolith aquifers are often saline, while on Karoo basalt, layers of fresh rock in banded weathering well digging is hard and thus favours screened regolith borehole construction.

37

Box 13: A review of current practice: Zambia

Zambia has experienced severe drought problems in recent years. As a result, food stocks have been depleted, livestock have died or been sold off to buy food, and water resources have generally declined. This decline has been particularly noticeable in the dry Southern Province of the country.

The geology of Southern Province is primarily crystalline basement rocks, basalt, alluvium and sandstone, with aquifers of low groundwater potential relative to other types. High spatial variability and low aquifer permeability make siting of productive water points difficult. However, a review of 1000 water point records found that:

- 141 existing water points (20%) are high-yielding and underutilized; these could immediately support production if pump capacity were increased above that of single handpumps fitted at present;
- 195 (27%) are low yielding because they are of inappropriate design in aquifers of low permeability; these water points should be converted to screened regolith boreholes, large-diameter wells or collector wells, which would then support production;
- 151 (21%) are low yielding because of inadequate siting; these water points could potentially benefit from re-siting using a combination of exploratory drilling and geophysical techniques developed for hard rock aquifers;
- 229 (32%) of existing water points are considered satisfactory for domestic supply.

Productive groundwater development has the potential to be of considerable benefit both to local people and the environment if existing water points were more effectively utilized and if new water points were chosen to suit the local aquifer conditions. It is conservatively estimated that 340 productive water points and associated small-scale irrigation schemes of total area 220 ha could immediately be developed to benefit about 85 000 people living in these driest parts of Zambia.

Groundwater in Zambia is recognized to be an underutilized resource, and massive investment is proposed in the national water resources master plan. Over 1000 water points have been constructed in Southern Province since 1995. Too many of these are low-yielding or dry, not because groundwater is particularly scarce but because the new water points have been inadequately sited or are of inappropriate design for the groundwater conditions. Also, the present investment is for domestic supply only. Siting and selecting appropriate well designs, and recognizing the opportunity for production at the outset rather than returning to successful water points, would be a more cost-effective approach to rural development and bring more immediate benefits to the local communities.

Box 14: The choice of appropriate technology

To be appropriate, a technology should:

○ be as inexpensive as possible without jeopardising the effectiveness of the improvements sought;

○ be easy to operate and maintain at village, community, or municipal level, and not demand a high level of technical skill or require a massive deployment of professional engineers;

○ rely on locally-produced materials rather than on externally provided equipment and spare parts, where this is practicable;

○ make effective use of local labour, especially in areas where there is a surplus of labour;

○ facilitate and encourage the local manufacture of equipment and parts under the leadership of entrepreneurs;

○ facilitate the participation of village communities in its operation and maintenance; and

○ be compatible with local values and preferences.

source: WHO (1987)

References and further reading

Acworth, R.A. (1987). The development of crystalline basement aquifers in a tropical environment. *Quarterly Journal of Engineering Geology*, 20: 265-272.

BGS (1989). The basement aquifer research project, 1984-1989: Final Report to the Overseas Development Administration. *Tech. Rept WD/89/15*, British Geological Survey, Keyworth, UK.

Bromley, J., Butterworth, J.A., Macdonald, D.M.J., Lovell, C.J., Mharapara, I. and Batchelor, C.H. (1999) Hydrological processes and water resources management in a dryland environment I: An introduction to the Romwe Catchment Study in Southern Zimbabwe. *Hydrology and Earth System Sciences* 3(3), 322-332.

Buckley, D.K. and Zeil, P. (1984). The character of fractured rock aquifers in eastern Botswana. In: Challenges in African Hydrology and Water Resources, Proceedings of the Harare Symposium, *IAHS Publ. No. 144*: 25-36.

Butterworth, J.A., Macdonald, D.M.J., Bromley, J., Simmonds, L.P., Lovell, C.J. and Mugabe, F. (1999) Hydrological processes and water resources management in a dryland environment III: Groundwater recharge and recession in a shallow weathered aquifer. *Hydrology and Earth System Sciences* 3(3), 345-352.

Butterworth, J.A., Schulze, R.E., Simmonds, L.P., Moriarty, P. and Mugabe, F. (1999c). Hydrological processes and water resources management in a dryland environment IV: Long-term groundwater level fluctuations due to variation in rainfall. *Hydrology and Earth System Sciences* 3(3), 353-361.

Chilton, P.J. and Foster, S.S.D. (1995). Hydrogeological characterisation and water supply potential of basement aquifers in tropical Africa. *Hydrogeology Journal* 3(1), 36-49.

Chilton, P.J. and Smith-Carington, A.K. (1984). Characteristics of the weathered basement aquifer in Malawi in relation to rural water supplies. Proceedings of the Harare Symposium: Challenges in African Hydrology and Water Resources, *IAHS Publication No. 144*, 57-72.

Clark, L. (1985). Ground water abstraction from basement complex areas of Africa. *Quarterly Journal of Engineering Geology*, 18: 25-34.

Commonwealth Scientific Council (1987). Groundwater exploration and development in crystalline basement aquifers. Proc. Workshop, Harare. June 1987. *Technical Paper 273. Volumes 1 and 2.*

Dev Burman, G.K. and Das, P.K. (1990). Groundwater exploration in hard rock terrain: an experience from eastern India. Proc. Beijing Symposium on: The hydrological basis for water resources management. *IAHS Publ. No. 197*, 19-30.

Foster S.S.D. (1988). Quantification of groundwater recharge in arid regions – a practical view for resource development and management. *NATO ASI Series C222*, 323-338. Reidel Publ., Dordecht.

Houston, J.F.T. (1988). Rainfall–runoff –recharge relationships in the basement rocks of Zimbabwe. I. Simmers (ed.), *Estimation of Natural Groundwater Recharge*, pp.349-365, Reidel Publ. Co.

Houston, J.F.T. and Lewis, R.T. (1989,). The Victoria Province Drought Relief Project, II. Borehole Yield Relationships. *Ground Water*, 26: 419-426.

Howard, K.W.F, Hughes, M., Charlesworth, D.L. and Ngobi, G. (1992). Hydrogeological evaluation of fracture permeability in crystalline basement aquifers of Uganda. *App. Hydrogeol.* 1, 55-65.

Howard, K.W.F. and Karundu, J. (1992). Constraints on the exploitation of basement aquifers in East Africa – water balance implications and the role of the regolith. *J. Hydrology* 139, 183-196.

Johnston, C.D. (1987). Preferred water flow and localised recharge in a variable regolith. *J. Hydrology*, 94, 129-142.

Lovell, C.J., Batchelor, C.H., Waughray, D.K., Semple, A.J., Mazhangara, E., Mtetwa, G., Murata, M., Brown, M.W., Dube, T., Thompson, D.M., Chilton, P.J., Macdonald, D.M.J., Conyers, D. and Mugweni, O. (1996). Small scale irrigation using collector wells: Pilot Project – Zimbabwe. Final Report. Oct 1992 - Jan 1996. *ODA Report 95/14*, Institute of Hydrology, Wallingford, UK, 110pp.

Lovell, C.J., Butterworth, J.A., Moriarty, P.B., Bromley, J., Batchelor, C.H., Mharapara, I., Mugabe, F.T., Mtetwa, G., Dube, T. and Simmonds, L. (1998). The Romwe Catchment Study, Zimbabwe: The effects of changing rainfall and land use on recharge to crystalline basement aquifers, and the implications for rural water supply and small-scale irrigation. Final Report to DFID, *Report 98/3*, Institute of Hydrology, Wallingford, UK.

Macdonald, D.M.J., Thompson, D.M.E. and Herbert, R. (1995). Sustainability of yield from wells and boreholes in crystalline basement aquifers. *Technical Report WC/95/50*, British Geological Survey, Wallingford, UK.

MacDonald, A.M. and Macdonald, D.M.J. (1997). Simple modelling to illustrate the impact of drought on groundwater availability. *Technical Report WC/97/1*, British Geological Survey, Wallingford, UK.

Margat, J. (1991). Les eaux souterraines dans le monde. Similitudes et differénces (The groundwaters of the world: similarities and differences). *Proc. 21 Journées de l'Hydraulique*, 29-31 January, Sophia Antipolis, France. vol IV, 1-13.

Morgan, P. (1990). *Rural Water Supplies and Sanitation*. Macmillan Publishers, London, UK.

Moriarty, P.B. and Lovell, C.J. (1998). Groundwater availability and environmental recovery during wet and dry cycles in Zimbabwe. *Proc. International Conference on*

Water Resources Variability in Africa during the 20th century. Abidjan, Cote d'Ivoire, Nov 1998, 105-108.

Nichols, W.D. (1993). Estimating annual groundwater discharge by greasewood in areas of shallow groundwater in the northern Great Basin using an energy-combination model. *Water Resources Research* 29, 2771-2778.

Nichols, W.D. (1994). Groundwater discharge by phreatophyte shrubs in the Great Basin as related to depth to groundwater. *Water Resources Research* 30 (12), 3265-3274.

Raugarajan, R. and Athavale, R.N. (2000). Annual replenishable groundwater potential of India – an estimate based on injected tritium studies. *J. Hydrology,* 234, 38-53.

Sophocleous, M. (1992). Groundwater recharge estimation and regionalization: the Great Bend Prairie of central Kansas and its recharge statistics. *J. Hydrology,* 137, 113-140.

Taylor, R.G. and Howard, K.W.F. (1999). Lithological evidence for the evolution of weathered mantles in Uganda by tectonically controlled cycles of deep weathering and stripping. *Catena* 35, 65-94.

Taylor, R.G. and Howard, K.W.F. (1996). Groundwater recharge in the Victoria Nile basin of east Africa: support for the soil-moisture balance approach using stable isotope tracers and flow modelling. *J. Hydrology* 180, 31-53.

UNESCO (1984). Ground water in hard rocks. Studies and reports in hydrology. *Report No. 33*, Paris.

Vegter, J.R. (1995). *An Explanation of a Set of National Groundwater Maps*. Water Research Commission, Pretoria, South Africa.

Wittenberg, H. and Sivapalan, M. (1999). Watershed groundwater balance estimation using streamflow recession analysis and baseflow separation. *J. Hydrology* 219, 20-33.

Waterlines (1998). Water source selection. *Technical Brief No. 55, vol. 16* (3), 15-18.

White, C.C., Houston, J.F.T. and Barker, R.D. (1988). The Victoria Province Drought Relief Project, I. Geophysical siting of boreholes. *Ground Water* 26 (3) 309-316.

WHO (1987). Technology for water supply and sanitation in developing countries. *Technical Report Series* 742, World Health Organization, Geneva.

Wright, E.P. and Burgess, W.G. (eds) (1992). The Hydrogeology of Crystalline Basement Aquifers in Africa. *The Geological Society Special Publication No.66*, London.

Wright, E.P, Herbert, R., Murray, K.H., Ball, D., Carruthers, R.M., McFarlane, M.J. and Kitching, R. (1989). Final report of the collector well project 1983-1989. *British Geological Survey Technical Report WD/88/3*, Keyworth, UK.

41

3 How much does a productive water point cost?

Introduction

FINANCIAL AND ECONOMIC ISSUES are critical to the successful introduction of productive water points. It may appear that the initial costs of productive water points are high compared to other water supply systems. However, it is important to take into account true pricing mechanisms and the wide range of benefits, some not easy to quantify in direct monetary terms. Apart from systems cost and cost/benefit analysis other aspects need to be considered, including affordability, willingness to pay, provision of subsidies and financing mechanisms, and the longer-term contributions that productive water points make to the local economy and local environment through diversification of livelihood strategies and intensification of dryland farming systems. A comment by Dijon (1987) is relevant in this context:

> ...the supply of water to communities on the edge of survival and settled on the fringes of deserts cannot be weighed only in economic terms. ... If the provision of basic water supplies helps to maintain the population in its ancestral lands, where some local resources such as arable soil and rangeland are still available, then a massive exodus, uprooting and destitution under subwelfare conditions can be avoided or at least slowed down: large tracts of land will not be abandoned. Eventually, some economic benefits will result, and major losses will be avoided.

No system for estimating costs ever provides 100 per cent accuracy, and with the multiple types of costs and benefits of water development projects in least-developed countries, far-reaching conclusions can rarely be drawn in strict economic terms and from a global perspective. For the same type of installation, for example, costs of groundwater development will be lower in areas with favourable hydrogeological conditions and shallow water tables, where materials and skills are available locally, and where greater inputs are available from the user community. On the other hand, in arid areas underlain by hard rock, with dispersed population settlements and little infrastructure, such costs may be high. Likewise, the decision on whether to use surface or groundwater or a combination of the two has physical, economic and social dimensions. Often surface water is preferred by users: it may be the traditional water source; people are used to its taste; it is free. But in some areas, surface

42

water may not be available at all, in others it may be contaminated or unreliable. With the related problems of population growth, overuse and contamination of surface water, a switch to either groundwater or a combination of surface and groundwater for water supply may be necessary and indeed desirable, even though more expensive in some settings.

Cost estimates quoted in the literature vary widely depending on the methods used to arrive at the figures, and the methods used are always open to debate. This again makes regional comparisons difficult. Factors that contribute to these differences include:

○ exclusion or inclusion of recurring costs, overhead project costs, personnel costs or transport costs;
○ exclusion or inclusion of opportunity costs of labour and local materials;
○ inflation and exchange rate fluctuations since the time of calculation;
○ different pricing in different countries due to import duties, devaluation of currencies etc.

The total cost of a given surface or groundwater project is comprised of capital and recurrent costs. Capital costs include the initial costs of exploration, data collection, scheme design, social development and construction, while recurrent costs include those for energy, labour, operation and maintenance, depreciation and interest charges. Ideally, all costs should be included and amortized over time to get a clear view of the implications of choosing different technologies. In particular, recurring costs in labour and money can vary widely between different technology choices and are important to long-term sustainability.

The following analysis of 13 productive water point options in Zimbabwe can therefore only provide a means of estimating future trends in other geographical areas. An up-to-date breakdown of costs should be made locally for each technology option before any decisions on the basis of cost differentials are made.

Cost/benefit analysis of 13 productive water point options

Thirteen ways to supply water and create a productive water point are:
1) Up-grade an underused small dam
2) Up-grade an underused borehole or well
3) Convert an existing borehole or well to a screened regolith borehole
4) Convert an existing borehole or well to a large-diameter well
5) Convert an existing borehole or well to a collector well
6) Site and construct a new small dam
7) Site and construct a new screened regolith borehole
8) Site and construct a new large-diameter well
9) Site and construct a new collector well

10) Hydrofracture an existing borehole
11) Site and construct a conventional deep borehole
12) Site and construct a deep well (or multiple deep wells)
13) Site and construct a family well (or multiple family wells)

Appendix 1 provides a breakdown of the capital costs, recurrent costs and economic viability of each option, with and without the opportunity cost of labour included, for both manual and mechanized pump technologies, and compared to a standard domestic borehole at which there is no production. Table 3.1 summarizes the results for the case where opportunity cost of labour for production is included and manual pump technology is used. A question that has been asked is whether multiple 'low cost' family wells offer a better solution than, say, single more expensive (but higher yielding) community wells. The performance indicators shown in Table 3.1 and Appendix 1 do not change and apply equally to both single and multiple units.

The following notes accompany the full analysis:

○ Capital costs are based on prices recorded in various water projects in 1994 and 1996, increased by a CPI inflator to December 1998 prices. Between 1994 and 1998 the Z$ depreciated substantially in real terms, meaning that technologies using imported equipment (such as collector wells) have become less attractive compared with systems using local equipment. Hence, the costs of these technologies have also been increased by a real devaluation inflator. Lastly, the capital costs of the 'new' productive water point options tested under pilot conditions have been increased by 10 per cent to account for the management costs and profits of the entrepreneur.

○ Inflators and exchange rates used to convert all Z$ figures to December 1998 US$ are:

Year	Inflation %	Index	Multiplier to 1998	Z$/day	Exchange rate	US$/day
1990	na	100.00	7.5756	1.13	2.9737	0.38
1991	23.30	123.30	6.1440	1.40	3.7393	0.37
1992	42.09	175.20	4.3240	1.99	5.1121	0.39
1993	27.63	223.60	3.3880	2.54	6.5289	0.39
1994	22.27	273.40	2.7709	3.10	8.1279	0.38
1995	35.15	369.50	2.0502	4.19	8.6654	0.48
1996	16.40	430.10	1.7614	4.88	10.0240	0.49
1997	20.07	516.40	1.4670	5.86	12.2300	0.48
1998	46.70	757.56	1.0000	8.59	37.3600	0.23

○ Yields of water ($m^3\ d^{-1}$) are average values recorded in the various projects (Table 2.4). Family well yields were measured by bucket-count at 110 sites before and after deepening in the Bikita Emergency

44

Water Supply Programme (WaterAid, 1994). Standard borehole yields were measured by a five-hour pumping test at 149 sites in the World Bank Drought Relief Programme in Masvingo Province. An average yield increase of 56 per cent was measured by pumping tests at 36 sites before and after hydrofracturing in a 1996 UNICEF Borehole Rehabilitation Programme. Five-hour pumping tests and modelling were used to project the sustainable yield during drought for screened regolith boreholes, large-diameter wells and collector wells. A conservative yield of 44 m^3 d^{-1} is used in the case 'upgrade underused borehole'; in practice, yields greater than 100 m^3 d^{-1} have been measured on some boreholes.

○ Pump types are as shown. A 10-hour pumping day is assumed. For small dams, the manual option uses three treadle pumps at US$150 and delivering 1 l s^{-1} each, the motorized option uses one petrol 2-5 HP low-lift centrifugal pump at US$900 delivering 3 l s^{-1}. In the case of groundwater, pump cost increases and pump capacity decreases with depth. For boreholes, the manual option is restricted to pump designs that can be fitted in multiples to a slim casing: hydropumps are used at US$800 each on a 180 mm diameter casing delivering 0.3 l s^{-1} each at 40 m head. On wells where space is not a constraint the manual option uses multiple Zimbabwe B-type bushpumps with 75 mm cylinders at US$245 delivering 0.5 l s^{-1} at 15 m head, and for the standard domestic borehole a single bushpump at US$490 delivering 0.3 l s^{-1} at 40 m head. The family well uses a Blair bucket-pump at US$171 delivering 0.1 l s^{-1} at 10 m head. The motorized option in each case uses one diesel-driven monopump at US$2300 delivering greater than 1.25 l s^{-1} at 40 m head.

○ Fenced area (ha) is calculated as 0.032 × Yield of water (m^3 d^{-1}) (see Chapter 4).

○ The average garden area supported by small dams is reported to be 3.2 ha and will require a yield of water of 100 m^3 d^{-1} applying the above empirical equation.

○ Land for the garden is valued at US$100/ha in this example.

○ Social development inputs provided by NGOs and government staff are assumed equal for surface water and groundwater projects alike and are valued at US$1000 per scheme.

○ Recurrent costs are based on depreciation, O&M, fuel and labour. All water points and gardens are assumed to have a life of 20 years; depreciation is calculated at 5 per cent per annum. All pumps are assumed to have a life of ten years; depreciation is calculated at 10 per cent per annum, with all pumps replaced in year 11.

○ O&M is calculated at 5 per cent per annum of pump capital cost.

○ Motor pump running costs, principally for fuel, are calculated at US$0.05 per m^3 of water pumped for an eight-month (240 day) growing season.

○ Opportunity cost of labour is defined as the best forgone alternative. It

Table 3.1 The economic viability of 13 productive water point options and comparison with a standard domestic borehole. The opportunity cost of labour for production is included, including water lifting and distribution, and manual pump technologies are used.

Option	Capital cost US$	Recurrent cost US$	IRR* (%)	NPV* (13%)	US$/m³ of water	US$/ha irrigation	Per capita capital cost US$	Benefit to cost ratio
Upgrade underused small dam	18 229	5047	22	17 888	0.89	6669	24.50	1.3
Upgrade underused well or b/hole	6034	2394	87	13 518	0.95	7095	8.11	1.3
Convert to screened regolith b/hole	7612	2158	33	6580	1.04	7825	10.23	1.1
Convert to large-diameter well	5921	1334	69	8669	0.95	7093	7.96	1.3
Convert to collector well	9082	1675	24	5006	1.02	7661	12.21	1.2
Site and construct new small dam	185 316	13 402	-ve	-297 063	2.65	19 846	249.08	0.5
New screened regolith borehole	9146	2235	21	3688	1.09	8157	12.29	1.1
New large-diameter well	7456	1411	32	5777	1.01	7575	10.02	1.2
New collector well	10 617	1751	17	2113	1.08	8071	14.27	1.1
Hydrofracture existing borehole	6089	2028	63	8773	1.01	7578	8.18	1.2
New conventional deep borehole	7419	1416	22	2389	1.13	8481	10.07	1.1
Deep well (multiple deep wells)	1199	378	∞	7730	0.93	7006	128.92	1.3
Family well (multiple family wells)	599	126	∞	6279	1.18	8864	64.45	1.0
Standard domestic borehole (no production)	4719	285	-ve	-9762	n/a	n/a	6.34	n/a

is assumed here to be income from rainfed crops. It was estimated that 50 per cent of households in the project area in 1990 had annual incomes below US$166. Of this, off-farm income accounted for 16 per cent leaving on-farm income US$139 or US$0.38 per day, equivalent to Z$8.59 or US$0.23 per day at December 1998 prices. It is debatable, however, whether rural poor, and women in particular, have an opportunity to earn an alternative income in the dry winter months in Zimbabwe, and whether opportunity cost of labour is in fact appropriate in this analysis. Performance indicators with and without opportunity cost of labour are shown.

○ Traditional irrigated horticulture is a highly labour-intensive activity. The average time recorded on garden-related tasks at groundwater-based schemes is 282 hours/year/household, comprising 240 hours on manual water lifting, distribution and pump repair, and 42 hours on land preparation, transplanting, weeding, harvesting and marketing. The average garden size is 0.5 ha with an average of 80 households per garden giving a total labour input of 45 000 hours/year/ha or US$1294/ha at US$0.23 per eight-hour working day. Where a motor pump is used, it is assumed that half the time (120 hours) spent on manual operations is still required for water distribution, plus the 42 hours for marketing, etc. giving a reduced labour input of 27 000 hours/year/ha or US$776/ha. It should be noted, however, that these labour inputs are higher than reported elsewhere. In West Africa, for example, horticulturists using manual water lifting and distribution means are reported to average 14 000 hours for the irrigation of one hectare over an equivalent eight-month growing season.

○ The gross margin of Z$82 751 per hectare or US$2215 at December 1998 prices is the long-term average recorded across nine groundwater-based schemes over seven years.

○ Given the various sources of capital available for this type of development, the total cost of the project to the community is expressed as total fixed cost plus total recurrent cost for 20 years without interest, to allow decision-makers to apply their own (appropriate) interest rates to calculate the annual capital charge.

○ Thirteen per cent inflation rate is used in this example. It is the average rate for selected arid zones (Tanzania, Chad, Rwanda, Niger, Mali, Kenya, CAR and Zimbabwe). It is likely that there will be fluctuations in both inflation and exchange rates, and this will have an effect on final Net Present Value (NPV) figures, but in the example a constant rate of 13 per cent is used for simplicity.

○ Thirteen per cent discount rate is used in this example to calculate Net Present Value figures.

○ The average population of member households at the groundwater-based schemes is 744.

○ The internal rates of return of productive water points (excluding small

47

dams) can be raised by the value to the local population of the domestic water supply provided. The domestic water benefits are accounted for by a 'partial budget analysis'. The cashflow of a standard domestic borehole is subtracted from the cashflow of the productive water point options that provide domestic water and irrigation water. The unquantified benefits from domestic water are thus netted out, so that the incremental costs of each option are being compared with the incremental benefits of production. The resulting 'partial IRR' does not tell us whether the productive water point is economically viable, because it does not include the benefits from domestic water, but it does tell us whether it is more or less viable than a standard borehole (positive figures indicating more viable).

Garden gross margins

Garden gross margins are an important decision variable. Operational data have been collected monthly at nine groundwater-based productive water points in Zimbabwe since 1992. As new schemes have come on line, the secretaries of each committee elected by the local people have kept records of inputs to, and outputs from, their gardens, including information on pump repairs, cropping patterns, irrigation, crop protection, fertilization, crop yields and market prices. Table 3.2 is an example of data collected by the Muzondidya community during 1998. The monthly visits by project staff to collect the data have also allowed insight into the social dynamics at each scheme and identification of management issues that are arising. These issues are discussed more fully in Chapter 7.

The communities' records show the financial performance of each scheme on a season-by-season basis. Table 3.3 is a summary of gross margins recorded at all schemes since 1992, with and without community labour costed. The figures are calculated using the 'garden gate' prices and include the imputed value of vegetables used by members for home consumption. Highly variable performance from scheme to scheme is a feature, with some marked fluctuations even at individual schemes from year to year. In most cases, low figures can be linked to temporary failure of leadership (and a corresponding increase in problems associated with communal tasks such as timely pump repair or establishment of nursery beds) or to specific problems such as pest and disease control (including plagues of mice), heavy frost which reduced returns in 1993 and 1994, and theft of fencing that allowed livestock to destroy crops. Under-reporting is also a problem with some communities providing incomplete records. Some notable returns are those recorded at Romwe during the severe drought of 1991/92, when all surface water sources in the region failed and the vegetables grown sold for high prices. Also, the high returns at Gokota, Dekeza and Matedze in their first seasons of operation, which indicate that within some communities there exist the necessary qualities of leadership, experience of gardening and

48

institutional structures to make community projects successful as soon as the primary constraint of water is removed.

Returns at all schemes over the seven years are generally high relative to other land use options in these dry areas. Gross margins ranging up to Z$109 000 per hectare (US$2900) with labour costed, and 155 000 Z$/ha (US$4150) without labour costed, illustrate the returns possible from small, intensively cultivated pieces of land when a reliable source of irrigation water is made available. There is growing evidence from studies elsewhere that these figures are not unusual, and that community-based irrigation is financially viable in spite of relatively low levels of capital investment. Meinzen-Dick et al. (1994), Perry (1997) and Rukuni (1997) are some among the growing literature demonstrating this economic viability and promoting community-owned small-scale irrigation as one of the most important tools that Africans should give priority to over the next few decades.

Figure 3.1 shows that there is nothing economically amiss with small-holdings and that community systems out-perform larger government systems in Zimbabwe. The value of production per unit area and per unit of water both increase substantially as plot size decreases.

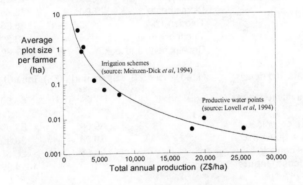

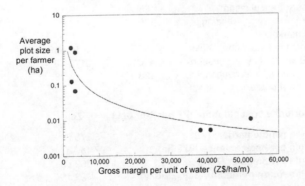

Figure 3.1 The relationships between size of irrigated land holding, productivity and water use efficiency (source: Lovell et al., 1996)

Table 3.2   Garden inputs and outputs recorded by the community at Muzondidya

Productive Water Point:	Muzondidya
Year or season:	1998
Membership (households):	134
Area of fenced garden (ha):	0.5
Opp. cost of labour (Z$ per 8 hr day):	8.59

Crop	Production (kg)	Price (Z$/kg)	Value (Z$)
Tomato	208	4.00	832
Rape	2397	3.00	7191
Sweet cabbage	3401	3.00	10 203
Maize	10 459 cobs	3.00/cob	31 377
Total value of output			**49 603**

Variable (cash) inputs	Type	Quantity (g)	Value (Z$)
Seed	Tomato	100	165.45
	Rape	500	164.55
	Sweet cabbage	480	190.00
	Maize	5000	62.00
	Cucumber	25	46.00
Pesticide	Diomethoate	500	78.75
	Thioden 1%	2000	35.65
Other	Transport, food		53.00
	Receipt book and lock		67.88

Pump repair (date and details)	Hired labour (Z$)	Spare parts (Z$)	Total Charge (Z$)
23/05/98 community repaired outlet pipe	55		55.00
20/08/98 outlet pipe welded		39.60	39.60
28/08/98 troughs repaired		39.95	39.95
01/09/98 3 bolts bought		26.55	26.55
Total variable (cash) inputs			**1024.38**

Variable (labour) input costs/household	Man-hours	Labour days (8 hr)	Value (Z$)
Land prep, nursery, transplanting	25	3.13	26.84
Irrigation pumping, watering, weeding	105	13.13	112.74
Harvesting and marketing		0.00	0.00
Total garden labour input per household	130	16.25	139.59
Total garden labour input by all households	17 420	2177.50	18 704.73
Total pump repair input by all households	123	15.38	132.07
Total variable input costs			**19 861.18**

Assumed opportunity cost of labour (Z$ per 8-hr day):	Zero	8.59
Gross margin (whole garden) (Z$)	48 578.62	29 741.82
Gross margin per hectare fenced (Z$/ha)	97 157.24	59 483.65
Gross margin per member (Z$/member)	362.53	221.95
Gross margin return per labour day (Z$/day)	22.15	13.56

Table 3.3 Garden performance over seven years, with and without labour costed

Gross Margin Z$/ha

	1992	1993	1994	1995	1996	1997	1998
Opp. cost of labour Z$/day	1.99	2.54	3.10	4.19	4.88	5.86	8.59
With labour costed							
Romwe	27 840	12 194	8100	-6677	-16 525	12 101	108 961
Muzondidya		4556	7821		-37 027	35 678	59 484
Gokota			3021	15 215	769	-1911	44 094
Dekeza			17 413	27 869	43 820	1052	90 345
Nemauka			-9908	33 983	7282	-16 134	57 187
Mawadze			1749	34 209	89 121	59 038	72 010
Matedze				73 887	-37 778	4196	43 442
Machoka					-556	NR	74 819
Masekesa					-21 675	-10 188	-12 233
Without labour costed							
Romwe	42 993	26 298	18 903	6150	18 221	37 109	129 795
Muzondidya		15 628	27 673	38 804	18 431	98 036	97 157
Gokota			35 348	64 956	47 476	29 449	63 001
Dekeza			47 459	62 423	76 245	29 705	154 983
Nemauka			2221	67 426	44 237	7975	78 144
Mawadze			11 183	52 264	112 240	85 626	104 465
Matedze				101 050	6965	21 148	73 342
Machoka					17 018	NR	91 978
Masekesa					20 012	15 340	41 730

Gross Margin Z$/member

	1992	1993	1994	1995	1996	1997	1998
Opp. cost of labour Z$/day	1.99	2.54	3.10	4.19	4.88	5.86	8.59
With labour costed							
Romwe	174	94	84	-65	-165	121	1090
Muzondidya		17	31	29	-137	133	222
Gokota			13	70	4	-10	237
Dekeza			178	279	421	10	836
Nemauka			-59	200	43	-119	340
Mawadze			17	342	857	568	720
Matedze				425	-217	24	250
Machoka					-3	NR	440
Masekesa					-128	-60	-70
Without labour costed							
Romwe	269	204	182	12	182	371	1298
Muzondidya		62	100	140	68	366	363
Gokota			137	256	247	157	339
Dekeza			383	503	733	286	1435
Nemauka			9	309	263	59	465
Mawadze			77	455	1079	823	1045
Matedze				511	40	122	422
Machoka					99	NR	541
Masekesa					118	90	240

Gross Margin Z$/labour day

	1992	1993	1994	1995	1996	1997	1998
Opp. cost of labour Z$/day	1.99	2.54	3.10	4.19	4.88	5.86	8.59
With labour costed							
Romwe	3.66	2.17	2.67	-3.54	-2.32	2.84	44.93
Muzondidya		0.96	1.42	1.11	-3.26	3.37	13.56
Gokota			0.34	1.57	0.08	-0.36	20.03
Dekeza			2.68	5.20	6.59	0.22	12.01
Nemauka			-2.68	7.65	0.96	-3.92	23.44
Mawadze			0.91	12.73	18.81	13.01	19.06
Matedze				20.59	-4.12	1.45	12.48
Machoka					-0.15	NR	37.46
Masekesa					-2.54	-2.34	-1.95
Without labour costed							
Romwe	5.65	6.12	5.77	0.65	2.56	8.70	53.52
Muzondidya		3.51	4.52	5.30	1.62	9.23	22.15
Gokota			3.44	5.76	4.96	5.50	28.62
Dekeza			5.77	9.39	11.47	6.08	20.60
Nemauka			0.42	11.84	5.84	1.94	32.03
Mawadze			4.01	16.92	23.69	18.87	27.65
Matedze				24.78	0.76	7.31	21.07
Machoka					4.73	NR	46.05
Masekesa					2.34	3.52	6.64

Benefit-to-cost ratios

Table 3.4 shows benefit-to-cost ratios of the various productive water point options at low, average and high values of garden gross margin. Table 3.5 ranks the options, with and without community labour costed. The ranking is based on ratios expressed as the sum of benefits achieved at each gross margin

Table 3.4 Benefit-to-cost ratios of productive water point options, with and without community labour costed

	Low Gross margin 1500 US$/ha/year		Average Gross margin 2215 US$/ha/year		High Gross margin 4000 US$/ha/year	
	with	without	with	without	with	without
1) Upgrade underused small dam						
a) multiple manual pumps	0.9	4.2	1.3	6.3	2.4	11.3
b) motor pump	1.0	2.0	1.5	2.9	2.6	5.3
2) Upgrade underused borehole or well						
a) multiple manual pumps	0.9	3.3	1.3	4.8	2.3	8.7
b) motor pump	1.0	2.1	1.5	3.0	2.7	5.5
3) Convert to screened regolith borehole						
a) multiple manual pumps	0.8	2.3	1.1	3.5	2.1	6.3
b) motor pump	0.9	1.7	1.3	2.5	2.4	4.5
4) Convert to large-diameter well						
a) multiple manual pumps	0.9	3.3	1.3	4.8	2.3	8.7
b) motor pump	0.8	1.3	1.1	1.8	2.0	3.3
5) Convert to collector well						
a) multiple manual pumps	0.8	2.5	1.2	3.7	2.1	6.7
b) motor pump	0.7	1.2	1.1	1.7	1.9	3.1
6) Site and construct new small dam						
a) multiple manual pumps	0.3	0.4	0.5	0.6	0.8	1.1
b) motor pump	0.3	0.4	0.5	0.6	0.8	1.0
7) New screened regolith borehole						
a) multiple manual pumps	0.7	2.1	1.1	3.0	12.0	5.5
b) motor pump	0.9	1.5	1.3	2.3	2.3	4.1
8) New large-diameter well						
a) multiple manual pumps	0.8	2.6	1.2	3.8	2.1	6.9
b) motor pump	0.7	1.1	1.1	1.7	1.9	3.0
9) New collector well						
a) multiple manual pumps	0.8	2.1	1.1	3.2	2.0	5.7
b) motor pump	0.7	1.1	1.0	1.6	1.8	2.9
10) Hydrofracture existing borehole						
a) multiple manual pumps	0.8	2.6	1.2	3.8	2.1	6.9
11) New conventional deep borehole						
a) multiple manual pumps	0.7	1.9	1.1	2.8	1.9	5.0
12) Deep well (multiple deep wells)						
a) manual pump	0.9	3.4	1.3	5.1	2.3	9.2
13) Family well (multiple family wells)						
a) manual pump	0.7	1.7	1.0	2.5	1.8	4.5

over the sum of costs. In terms of benefit-to-cost ratio, it is clear that the best water supply option depends on whether you value community labour or not, and if you do, then mechanized pump technology becomes increasingly important.

Table 3.5   Ranking of productive water point options in terms of benefit-to-cost ratio

Rank	Community labour costed	Rank	Community labour not costed
1	Upgrade underused b/hole, motor pump	1	Upgrade underused small dam, manual pumps
2	Upgrade underused small dam, motor pump	2	Deep well, manual pumps
3	Upgrade underused small dam, manual pumps	3	Convert to large-diameter well, manual pumps
4	Convert to screen regolith b/hole, motor pump	4	Upgrade underused b/hole, manual pumps
5	Deep well, manual pumps	5	New large-diameter well, manual pumps
6	New screened regolith b/hole, motor pump	6	Hydrofracture existing b/hole, manual pumps
7	Convert to large-diameter well, manual pumps	7	Convert to collector well, manual pumps
8	Upgrade underused b/hole, manual pumps	8	Convert to screened regolith b/hole, manual pumps
9	New large-diameter well, manual pumps	9	New collector well, manual pumps
10	Hydrofracture existing b/hole, manual pumps	10	New screened regolith b/hole, manual pumps
11	Convert to collector well, manual pumps	11	Upgrade underused b/hole, motor pump
12	Convert to screen regolith b/hole, manual pumps	12	Upgrade underused small dam, motor pump
13	Convert to large-diameter well, motor pump	13	New conventional deep borehole, manual pumps
14	New collector well, manual pumps	14	Convert to screen regolith b/hole, motor pump
15	New screened regolith b/hole, manual pumps	15	Family well, manual pump
16	Convert to collector well, motor pump	16	New screened regolith b/hole, motor pump
17	New large-diameter well, motor pump	17	Convert to large-diameter well, motor pump
18	New conventional deep borehole, manual pumps	18	Convert to collector well, motor pump
19	New collector well, motor pump	19	New large-diameter well, motor pump
20	Family well, manual pump	20	New collector well, motor pump
21	New small dam, motor pump	21	New small dam, manual pumps
22	New small dam, manual pumps	22	New small dam, motor pump

The economic benefit of more reliable water

At a productive water point, the combination of adequate pump capacity and a strong degree of community involvement in operation and maintenance means that water is quickly available and can be relied upon. Across the schemes in Zimbabwe, the supply has not dropped below 10 m³ per day, even at the height of drought, when demand on the water points is greatest.

53

Overall, 50 per cent of garden members and 39 per cent of non-members say they now obtain all their domestic water from the schemes, and on top of this, up to 54 per cent of those surveyed at each site say they now also use the productive water points for domestic water when their nearest other source fails. In contrast, many standard boreholes and wells in the area are found to be much more seasonal in nature, slow to yield water, and are often not quickly repaired when broken.

The economic importance of providing reliable water supplies has been recognized for some time and is often taken as more than enough economic justification for the implementation of a rural water supply project. Participatory surveys and contingent valuation studies at the productive water points have noted the high value attached to the reliability of the water supply. For example, respondents consistently ranked 'reliability' high on their list of perceived benefits, above 'closeness' of the schemes, and a random sample of garden members chose, on average, to walk 600 m further to a productive water point than to another water source (Table 1.2).

The economic benefit of improved food and income security

A reliable water supply, combined with a secure vegetable garden, means that the projected income from the garden has a high probability of being earned. As a result, household purchases and new projects can be planned in advance. The lowering of risk in household planning has an economic value, and is reflected in sharp increases in both the perceived value of allotments in the gardens and in their actual purchase price. A comment typical of many garden members is *'If I gave up my plot I'd be giving up my future'*. At a scheme where gross financial earnings per member were US$82 in 1995, the price for a plot of land in the garden rose from US$2 to US$26 between 1993 and 1995.

The economic benefit of community-based maintenance

It is critical that local communities are empowered to own rural water supplies and assume responsibility for operation and maintenance (O&M). This becomes possible when a water point assumes an economic value through production because this creates both the incentive and the income needed to make community-based pump maintenance a reality. At productive water points in Zimbabwe, pumps have been maintained locally from the outset or the users have chosen to pay government pump-minders to carry out repairs. At all schemes, it is the users who are choosing to buy spare parts (or pay pump-minders) using income generated by the schemes: they no longer rely on external assistance. Box 10 shows payment regimes recorded for O&M. Although these have been arbitrary, ranging from US$50-245 per year, they are not insignificant, and are in stark contrast to conventional rural water supply where the annual cost to the state (or donor) is of the order US$90, which equates to an additional discounted cost of US$1120 per domestic water point over a standard lifetime of 20 years.

Affordability, willingness to pay and cost recovery

Water supply systems will be sustainable only if sufficient resources are recovered to keep them operational. Covering only maintenance costs is not adequate. In the long run it is absolutely necessary to generate resources for replacement and investment. This depends on adequate attention being given at design to affordability, willingness to pay and cost recovery.

Community members often can and are willing to provide a considerable contribution in cash and kind to obtain a better water supply. Encouraging such contributions will help a government or donor to achieve a better coverage at lower cost. More important, it will stimulate users to accept greater responsibility for the system.

The contributions households are willing to pay will depend on the benefits they perceive and whether local ownership and tenure are secure. A review of water harvesting in African countries, for example, showed that contributions to construction varied from 10 to 40 per cent. Communities can be asked for their own estimates to get an impression of the willingness to pay. Since many of the benefits of productive water points will particularly concern women, their involvement in the decision-making process is vital.

A 1995 contingent valuation survey invited willingness-to-pay (WTP) responses from local people at six productive water points in Zimbabwe. In total, 60 households were surveyed, including both members and non-members of the associated gardens. Respondents were asked in discrete yes/no format about their WTP to create and join the irrigation scheme and to maintain the water point. In order to eliminate bias, an iterative bidding game using different opening bids as starting points was used. The survey was conducted in Shona by a female enumerator, and most respondents took the exercise very seriously, with lengthy discussions between husband and wife, or whichever female member took an interest in the garden. For the question about creating the scheme, where larger sums of money were involved, bags of maize were used as a proxy for cash. The garden plot prices that people were paying at the time when a space arose ranged from US$6 to US$33.

The survey found a 95 per cent confidence interval for a WTP a one-off joining fee to create the scheme of between US$25 and US$33. With an average of 80 households using each scheme, these figures indicate a potential for cost recovery of 33-44 per cent of the capital cost of up-grading an existing underused well or borehole, or 18-25 per cent of the cost of siting and constructing a more expensive new collector well where this is the necessary and appropriate technology choice. In practice, a payment vehicle of bags of maize could be set equivalent in value to the prearranged fee, the maize to be sold by the water point committee to raise the matching funds. Alternatively, a series of household payments totalling the identified WTP could offset the problem of access to lump sums of cash and perhaps be stratified for different income groups.

The survey found a mean WTP to maintain a productive water point of US$0.55 per month or US$6.60 per year per household. With an average of 80 households using each water point, rising to at least 100 at the end of each dry season, the survey suggests that over US$500 could be collected each year to cover O&M costs. In practice, O&M costs being paid range from US$50 to US$245 (Box 10). This suggests potential for excess funds to be saved to cover replacement costs or to be invested in fertilizer, pesticides, marketing strategies, new income-generating projects or other community-based amenities such as toilets.

If WTP for maintaining the productive water points is taken as the total economic value of more reliable water and improved food and income security, and discounted over the scheme's 20 year time horizon at 13 per cent, this value aggregates out at US$36 550, a significant economic benefit.

Subsidies

Design criteria which tend to increase the capital cost of a productive water point include:

○ the physical location – often difficult areas where reliable water is not readily available and where greater care is needed to site and select the most appropriate technology;

○ the target yield – at greater than 10 m^3/day of water this is higher than yields accepted in conventional domestic water supply;

○ the social, economic and institutional setting and associated development cost – poor communities in communally-managed areas who, by the end of scheme implementation, need to be empowered to own, manage and maintain the resource.

Given these criteria and the associated costs, it will often not be possible for communities to install productive water points without some external support. Covering a large portion of the cost by subsidies, however, entails a risk that communities will become more dependent and see the system as not belonging to themselves (Box 15). Striking a balance between subsidies and community contributions is a very important process that is best established in dialogue with the communities.

Each programme has to decide the extent of the subsidies, which groups should benefit, and under what conditions. Setting the conditions is sensitive. They must be suited to the local culture and to the local institutional and economic capacity to organize and make regular contributions to communal activities. Where users are required to provide financial contributions in cash or kind, often only the most prominent in the community can participate and

Box 15: Subsidies

Subsidies are a standard component of virtually all rural development projects. It is difficult to find examples of government or non-government projects that do not include substantial funding from the sponsoring agency. Such funding can take several forms, helping to pay for labour, agricultural inputs, machinery services or technical expertise. Sometimes assistance is provided to help rural people carry out work on their own, and sometimes the work is done for them.

Subsidies for water resource development are intended to support rural development and improve human welfare simultaneously, but combining them can actually undermine the efforts. In economic terms, introducing a subsidy is justified only if two broad conditions apply: (1) there must be a market failure; and (2) a subsidy must be the best way to correct the market failure.

Subsidies used inappropriately can have many drawbacks. Heavy subsidies guide farmers to accept technologies they may not want and will not maintain, inhibiting project managers from gaining feedback on what farmers actually want. Financial subsidies can discriminate against products and practices that are not subsidized, thus impeding scientific progress and stifling indigenous knowledge. Subsidies can also undermine people's incentives to take the initiative, and as a result, rural development agencies promoting self reliance find it difficult to operate without subsidies because villagers become accustomed to 'give-aways'. There are numerous examples around the world to demonstrate these problems.

Alternative approaches to water resource development with no or low direct subsidies include education to allow people to do for themselves, supporting community organizations to rejuvenate collective action instead of subsidizing technology, institutional innovations to manage local externalities and spread the benefits (such as formal or informal compensation mechanisms whereby a group that benefits from a collective management arrangement secures co-operation from a group that does not), granting and specifying long-term property rights, improving credit and insurance markets, and sequencing projects to reduce financial constraints.

Evidence from around the world shows that farmers will invest in development when it is profitable for them to do so. This suggests that farmers do not need subsidies so much as they need less expensive, more profitable technologies; policies that encourage them to take a long-term perspective in caring for their land; greater awareness of available options; and support to organize themselves to make this investment.

Adapted from Kerr et al. (1996)

subsidies intended to provide water for all can end up benefiting only wealthier individuals. This highlights the need for differential subsidies in some cases and concerted effort at design to ensure that the poorer groups also have

57

access to the improved water supply and are not being marginalized further This aspect is discussed in Chapter 6.

Financing mechanisms

The successful provision of services including water to poor people living in dryland areas is a growing challenge. Far from improving coverage and raising standards of living, conventional approaches to service delivery have failed to keep pace with increasing and changing demands. Investment in inappropriate facilities, coupled with inadequate arrangements for their operation and maintenance, have led to a situation of waste, where infrastructure rapidly falls to pieces. In response, new approaches to service provision widely recognize the role of consumers, even poor consumers, in decision-making and financing services. For many poor people, taking control of decision-making and increasing their financial stake in service provision is an attractive proposition. Small-scale credit services are often utilized by poor communities to access services that have an associated investment cost. Increasingly, financing agents and domestic financial organizations are seeking to channel more funds into the credit sector, and there is enormous potential in the partnerships that can be developed between donors, microfinance institutions, local government, technical agencies and communities.

A particular lesson that emerges, however, is the need to consider how to effectively channel the technical assistance required for capacity building to community groups and microfinance institutions (Box 16). Besides community action, strong linkages with local authorities are critical for successful infrastructure projects. There is a need to understand and detail the responsibilities of different stakeholders, clarify the different roles required, and match appropriate actors with these. These institutional arrangements are discussed in later chapters.

Lease agreements

One example of an appropriate institutional arrangement to help promote cost recovery and ensure operational technology is the pump lease concept. This is a formal contractual arrangement established between the community, the local authority and the commercial sector (Box 17). Productive water points extend the range of pump technologies that the users are able to pay for and maintain – the pump lease concept takes this further and makes efficient, high quality technology available to poor people at a reasonable price.

Economic decision support

In summary, productive water points may carry a higher initial capital cost than standard domestic water points but their benefits accrue over time, not

Box 16: The right to small-scale credit schemes

In the past five years, there has been a marked growth in the number and outreach of microfinance institutions disbursing infrastructure loans to the urban poor in India. This is largely due to an institutional shift to recognize, support and facilitate these credit organizations working at field level to enable poor communities to access improved infrastructure. The potential of small-scale credit schemes is universal, and there are important lessons for rural water supply.

❍ Access to credit is vital but access to technical and supervisory support is also vital. While the poor are willing to pay for services, they also require technical assistance and a liaison agency to interact with the various parties. In particular, the greatest need of poor communities involved in procuring improved infrastructure, including water supply, is for institutional support to provide a linkage with local government as the ultimate providers of infrastructure services.

❍ Microfinance has tremendous potential as an instrument for poverty reduction, but complementary efforts, such as teaching women to read and write and training, are necessary to help those who lack the skills to make credit work for them. Especially in the area of infrastructure development, where interaction with local authorities and other officials is required, capacity building, exposure and negotiation skills training for community leaders are very important. Support is also needed in terms of an easily accessible, low-cost extension service or resource centre to help with efficient procurement of raw materials, cost estimations, supervisory help with physical construction, and training for maintenance of the facilities.

❍ The linked issues of collateral and legal land tenure are contentious, especially when viewed with regard to the sanction of credit. The reality is that the majority of poor people live and work on public land and lack legal land tenure. Hence they are unable to provide any form of traditional collateral for loans. An assurance by local government that the development belongs to the people and that the people will not be relocated for a minimum of 10 years can help. This 'near legal' tenure status has proved a powerful incentive for poor communities to 'buy-in' to the project and pay their own contributions.

❍ Regular savings are critical to ensure high loan repayment rates. All successful schemes insist on regular savings over a sustained period (ranging up to one year) before a loan may be applied for. Thereafter, some micro-finance institutions use 'peer pressure' through the self-help group to ensure loan repayments. Others rely on field workers and close interaction and familiarity with the financial situation of the group, while community owner-managed financial institutions (e.g. revolving funds) themselves decide which loans are to be sanctioned.

❍ While the loans to finance service provision may be individual, collective action is essential to bring about actual improvements in infrastructure.

Adapted from Ghatate (1999)

59

Box 17: The pump lease concept

The pump lease concept is a demand responsive approach to water point maintenance that has been applied recently in West Africa but not widely elsewhere. It may be expected that many communities in other dryland areas will welcome the idea of being able to rent a pump with guaranteed maintenance for a reasonable price.

In the lease concept, a local water service company (or NGO) owns a range of pumps that are placed with individual communities upon their request, based on a clear contractual arrangement for maintenance that ensures adequate service, outlines tariffs and explains what will happen if maintenance is not provided or if the company ceases operation. The company guarantees the good functioning of the pump, and when needed, will repair it within a set number of days. The lease includes regular preventive maintenance and spares supply, and ideally maintenance of the water point e.g. borehole cleaning, every 5 to 10 years.

The pump remains the property of the company, although arrangements could be made for lease-and-buy. The water point remains the property of the community. In the case of an existing pump owned by the community from a previous project, the company may repair the pump using a contract that reflects the maintenance demands of the particular type, or may propose changing to a more reliable pump and more favourable tariff as a long-term solution.

The contribution from the user community will typically be of the order of US$1–2 per family per month. In remote or low-income areas, a subsidy by government may cover extra transport costs. Introducing the concept is helped if 'free' handpumps are no longer provided with new water points. Donors should instead help to train and establish the private companies or NGOs as the service provider in the pump lease concept.

Choice of pump technology should remain with the community, but the concept can make more expensive higher quality pumps available at a low cost and with maintenance guaranteed. This is important for productive water points where wear-and-tear on handpumps is higher than at domestic water points and where motorized pumps can have an important role to play in making effective use of the available water resource.

A caretaker for each pump or each community should be included in the contract to ensure local protection and good service and to collect monthly payments from the pump users. The caretaker will earn say US$15 per month, have a legal contract with the water service company, and must be accepted by the community. The lease concept/private sector approach to water point maintenance is one that should be explored. It will help to develop business opportunities for local firms who can earn a living by installing and maintaining leased pumps and who will help to bridge the gap between pump dealers and local people.

Source: Paul van Beers, pers. comm.

only via lower recurrent costs to the state but also in a range of extended benefits to the wider production system. When evaluating the merits of different water supply options, the temptation is to use a cost basis only. The recent work clearly shows that more emphasis should be placed on the benefit side of the equation, especially as demand assessment (identifying what the communities actually want from their water projects and what they are willing to pay) is a key issue that should underpin the design of future rural water supply programmes.

Costs of surface water and groundwater production presented in the literature show a considerable variation. Examples in Table 3.6 illustrate the range, from low-cost hand-drilled tubewells in Asia through to flash distillation in Europe. Although this table provides only a global indication, it shows that costs of productive water points are no higher than other water supply systems.

Table 3.6 Annual capital costs of surface and groundwater development (source: UNDTCD, 1987; Lee and Visscher, 1992)

Option	Country	Annual capital cost per m^3 of water (US$)
hand-drilled tubewells	Bangladesh	0.01-0.02
shallow wells	Kenya	0.06
rock catchments	Kenya	0.09
boreholes	Morocco	0.05-0.10
surface water	Morocco	0.10-0.35
sub-surface dams	Kenya	0.11
boreholes	Mozambique	0.11
ferrocement ground tanks	Kenya	0.12-0.27
productive water points	**Zimbabwe**	**0.19-0.38**
boreholes	Malta	0.22-0.25
ferrocement standing tanks	Kenya	0.22-0.49
surface water	Mozambique	0.26
reverse osmosis	Malta	0.75
flash distillation	Malta	1.17

At a local level, the economic analysis reveals a number of guiding principles for the development of productive water points in dryland areas:

o The importance of garden gross margins as a decision variable – only if these fall below about US$1500/ha/year do productive water point options become unviable. At average returns of US$2215/ha/year recorded in Zimbabwe all options (apart from new small dams) are viable and at gross margins up to US$4000/ha/year now being recorded at more progressive schemes the returns per member become highly attractive.

61

○ Pump technology has a significant impact on systems cost and economic performance. Multiple hand or foot pumps fitted to single water points will be important to make effective use of available yields, but there is need to develop low-cost user-friendly designs which can be fitted to slim boreholes to draw water from depth. Motorized pumps will also play an important role, particularly where community labour is valued, where water tables are deep and where water point yields are high. The choice of appropriate pump technology is detailed in Chapter 4.

○ The superiority of upgrading over new technologies. Underused, rehabilitated, deepened, converted and hydrofractured existing water points all make effective use of sunk costs spent previously locating and storing water, offer high internal rates of return in their own right, and offer higher benefits than a standard domestic borehole. It should be noted, however, that the rehabilitation of small dams is viable only at siltation rates less than 5 per cent per annum (Box 18), and that the hydrofracturing analysis assumes that improvement in yield recorded on low-yielding boreholes holds true for higher-yielding sites. Hydrofracturing may improve permeability in the vicinity; however, the storativity of the reservoir (regolith) supplying the borehole is not changed, and if it is limited the yield is still low. In this case, hydrofracturing may result in overdraft not allowing a sustainable yield.

○ Test drilling, although more expensive than conventional geophysical siting, is economically viable due to the reliable yields of water and associated production made possible by this new approach. Conventional boreholes sited by geophysics do not fare well in this analysis, even where pump capacity is matched to yield to allow production, and there will generally be 10–15 alternative options that are better, depending on local groundwater conditions.

○ Large-diameter wells and screened regolith boreholes sited by test drilling have an important role to play in crystalline basement aquifers, although the economic viability of the latter is presently constrained by the lack of user-friendly pump technologies that can pump sufficient water from depth.

○ More costly options such as collector wells will be the technology of choice only in ground conditions where other options fail so it is important to adopt an approach to siting and selection that investigates the other options first.

○ One has to be careful with the siting and construction of new small dams to support irrigation because of the high costs per unit of water stored; this is not to say that new small dams have no place. On the contrary, a new small dam may be all that can be made available in a particular location and can be of great benefit to users in deprived semi-arid regions (Box 18), but it is important from an economic point of view to consider other options first.

Box 18: Choosing between surface and groundwater development

Advantages of surface water (small dams):
- ○ resource conservation through flood control and enhanced groundwater recharge, and through local people instinctively seeing the need to protect the local catchment to reduce siltation of the dam
- ○ readily accessible, needing only low-lift pump technology and incurring lower operation and maintenance costs
- ○ livestock can be watered without need to pump water
- ○ can provide fresh fish, reeds etc.
- ○ can increase bio-diversity by providing a sanctuary for wildlife and birds

Advantages of groundwater:
- ○ suited to phased, incremental development
- ○ a more permanent supply less vulnerable to drought
- ○ can be developed at multiple sites better suited to settlement patterns and reducing the time and energy spent in water collection
- ○ can be developed in flat areas not suited to dams
- ○ can lower the water table, reducing salinity problems in crop production
- ○ less prone to contamination, generally providing a potable supply without need for treatment due to natural filtration as the water passes through the ground
- ○ no associated health hazards such as bilharzia, cholera or malaria

Siltation of small dams:
A disadvantage of small dams is that they can silt up quickly in dryland areas prone to overgrazing and erosion. Economic appraisal in Zimbabwe shows that at a discount rate of 12% and amortization over 20 years, the rehabilitation of small dams valued at US$210 000 and currently 30% silted is viable and gives an IRR of 17% if the siltation rate is reduced to 1% per annum. At a siltation rate of 2.5% the IRR falls to a marginal 5.4% and at a siltation rate of 5% per annum and above the rehabilitation of small dams ceases to be viable. This analysis stresses the importance of proper natural resource management by the community within the catchment of the dam.

Conjunctive development of surface water and groundwater:
In water-short areas it will become increasingly important to consider combined development of surface water and groundwater. An example of this is found in the driest parts of Zimbabwe where groundwater potential is low due to low rainfall and the difficulties of water point siting in areas of younger intrusive granite and older gneiss. In this environment, small dams hold sufficient runoff to enhance recharge locally. The most reliable wells and boreholes are invariably located immediately downstream, and as the dry season progresses these water points maintain supplies for small-scale irrigation and livestock after the dams go dry. There is scope to investigate the productive potential of collector wells sited downstream of existing small dams, drilling the lateral boreholes beneath the dams to exploit fully this locally enhanced groundwater resource.

○ Construction of multiple 'low-cost' family wells will generally not be the most cost-effective way to support irrigation at community level in dryland areas overlying hard rock aquifers. However, enlarging and deepening existing family wells to the base of the regolith is viable. A weakness in this analysis is the lack of rigorous pumping test data and monitoring, but if average deep well yields measured by bucket count ($7.1 \text{ m}^3 \text{ d}^{-1}$) are sustained during drought, then deep wells are a cost-effective way to provide communities with water.

○ Traditional irrigated horticulture is highly labour intensive. The average time recorded on garden-related tasks is 282 hours/year/household or 45 000 hours/year/ha. Time-saving interventions will be important and will relate principally to the choice of water lifting device and distribution method rather than to the irrigation method *per se*. A pump capable of lifting water to an elevated tank, for example, will cost more initially but will quickly be repaid through time saving achieved by gravity feed irrigation. If irrigation time is halved, two relatively expensive hydropumps and a tank costing US$1800, fitted to a new large-diameter well, will return a benefit-to-cost ratio of 1.4 instead of a ratio of 1.2 if two standard handpumps costing US$490 are used.

The following chapters consider the technical, social and institutional steps to develop productive water points, beginning with identification, appraisal and conversion of existing water points that are underutilized or of inappropriate design.

References and further reading

ADB (1999). *Handbook for the economic analysis of water supply projects*. Asian Development Bank, Manila, Philippines. 361 pp.

Brookshire, D.S. and Whittington, D. (1993). Water resources issues in developing countries. *Water Resources Research*, 29 (7) 1883-8.

Carter, R.C., Demessie, D. and Mehari, M. (1998). Not a numbers game – making policy for maximum impact and sustainability. *Waterlines* 16 (3) 21-23.

Carter, R.C., Tyrell, S.F. and Howsam, P. (1996b). Strategies for handpump water supply programmes in less-developed countries. *J. CIWEM* 10, 130-136.

Dijon, R. E. (1987). Some economic aspects of groundwater projects executed by the United Nations in developing countries. *Proceedings of a Symposium on Economic aspects of groundwater exploration and assessment, UNDTCD*, Barcelona 19-23 October, 49-56.

Dore, D. (1997). *Proposed small dams and community resources management project: an economic appraisal*. Department for International Development in Central Africa, Harare.

Foster, S.S.D. (1989). Economic consideration in groundwater resource evaluation. *Developments in Water Science* 39, 53-65.

Katco, T.S. (1990). Cost recovery in water supply in developing countries. *Water Resources Development* 6(2), 86-94.

Ghatate, S. (1999). *Credit Connections: Meeting the infrastructure needs of the informal sector through microfinance in urban India.* UNDP-World Bank Water and Sanitation Program - South Asia. New Delhi, India.

Jobin, W. (1999). *Dams and disease: ecological design and health impacts of large dams, canals and irrigation systems.* E & FN Spon, London. 580pp.

Kerr, J.M., Shanghi, N.K. and Sriramappa, G. (1996). Subsidies in watershed development projects in India: Distortions and opportunities. *Gatekeeper Series No. 61*, IIED

Lee, D. and Visscher, J.T. (1992). Water harvesting: A guide for planners and project managers. *Technical paper series No. 30.* IRC International Water and Sanitation Centre, The Hague, The Netherlands.

Less, C. and Andersen, N. (1993) Hydrofracture – state of the art in South Africa. *Memoirs of the XXIV Congress of IAH*, Oslo, 717-723.

Lovell, C.J., Nhunhama, G., Sunguro, S. and Mugweni, O. (1998c). An economic impact: productive waterpoints in dryland areas. *Waterlines* 17(2), 5-8.

Lovell, C.J., Murata, M., Brown, M.W., Batchelor, C.H., Thompson, D.M., Dube. T., Semple, A.J. and Chilton, P.J. (1994). Small scale irrigation using collector wells pilot project: Zimbabwe. Fourth Progress Report, April 1994-September 1994. *ODA Report 94/9.* Institute of Hydrology, Wallingford, UK. 91pp.

Meinzen-Dick, R., Sullins, M. and Makombe, G. (1994). Agro-economic performance of small-holder irrigation in Zimbabwe. In: *Irrigation performance in Zimbabwe.* (eds) M. Rukuni, M. Svendsen and R. Meinzen-Dick, pp. 63-88. Proceedings of a UZ/AGRITEX/IFPRI Workshop, University of Zimbabwe, Harare, August 4-6 1993.

Perry, E. (1997). Low-cost irrigation technologies for food security in sub-Saharan Africa. Proc. Sub-regional Workshop: Irrigation technology transfer in support of food security. *Water Reports No. 14*, FAO, Rome, 91-104.

Rukuni, M. (1997). Creating an enabling environment for the uptake of low-cost irrigation equipment by small-scale farmers. Proc. Sub-regional Workshop: Irrigation technology transfer in support of food security. *Water Report No. 14*, FAO, Rome, 35-56.

Schiffler, M. (1998). *The economics of groundwater management in arid countries: Theory, international experience and a case study of Jordan.* GDI Book Series No. 11, Frank Cass Publishers, London.

Stoner, R.F., Milne, D.M. and Lund, P.J. (1979). Economic design of wells. *Quart. J. Eng. Geol.*, 12, 63-78.

UNDTCD (1987). Economic aspects of groundwater exploration and assessment. *Proceedings of a Symposium*, Barcelona 19-23 October.

Waughray, D.K., Moran, D. and Lovell, C.J. (1997). Potential uses of contingent valuation in the evaluation of dryland resource development projects: a small-scale irrigation case study from south-east Zimbabwe. In: *Sustainable Development in a Developing World: Integrating socio-economic appraisal and environmental assessment.* Eds: C. Kirkpatrick and N. Lee, Edward Elgar Publ., Cheltenham, UK. pp 200-216.

Waughray D.K., Lovell C.J., Mazhangara E (1998). Developing basement aquifers to generate economic benefits: a case study from south east Zimbabwe. *World Development* 26 (10), 1908-1912.

World Bank (1993). The demand for water in rural areas: determinants and policy implications. *The World Bank Research Observer* 8(1), 47-70.

4 Making more effective use of existing water points

The general approach

LOW DRILLING SUCCESS RATES reflect the difficulty of siting reliable water points in dryland areas (Table 1.1). However, the sheer numbers of wells and boreholes constructed in recent decades through investment in rural water supply mean that invariably one or two reliable, well-sited water points will exist in most dryland areas. While it is perhaps more attractive to development agencies to site and construct new water points, a major finding of the work in southern Africa is the number of existing underutilized water points. This is often because present pump capacity (either a handpump or a bucket and windlass) is far less than the potential safe yield. Queues of people waiting for their turn to pump water (as opposed to waiting for the water level in the well or borehole to recover) is a good indication that pump capacity, rather than water point yield, is the constraint. Up-grading these water points offers the first and most cost-effective option for increased water supply and productive groundwater development in many areas.

A second major finding is the number of water points of inappropriate design (see Box 13). Many water points currently being implemented in rural water supply programmes are neither technically appropriate for the local groundwater conditions nor the most cost effective in terms of cost per unit of water produced. Generally, these are deep boreholes of small diameter sited in shallow regolith aquifers of low permeability. An example of the problems faced by local people when this happens is given in Box 19. By reviewing past records, it is possible to identify these viable regolith aquifers and locations where water points of more appropriate design may be expected to provide improved security of water supply and allow production.

Figure 4.1 shows the key steps in this approach to hydrogeological investigation that is designed to make full use of available information and local knowledge in order to:

O identify and upgrade existing, underutilized water points;
O identify existing water points of inappropriate design and convert these to designs more suited to the local groundwater conditions;
O site and select new water points of appropriate design, where this is necessary.

This chapter details the first part of this approach, up to conversion. The

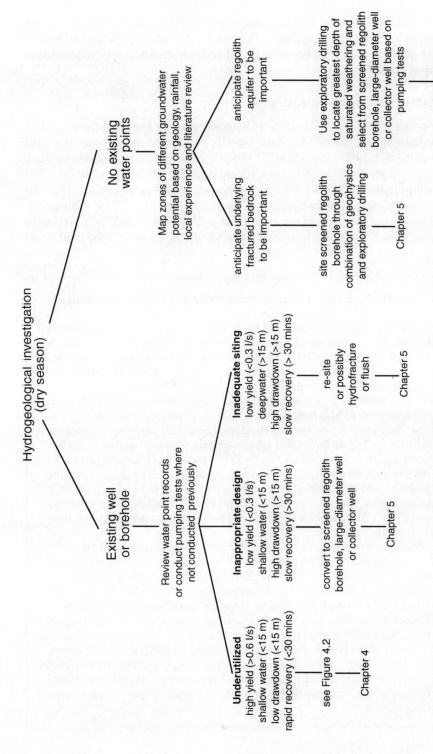

Figure 4.1 Key steps to site and select appropriate well designs in hard rock aquifers

steps taken to site and construct new water points of appropriate design are detailed in Chapter 5. Together, Chapters 4 and 5 form the technical part of a broader development plan, with social and institutional aspects detailed in Chapters 6 and 7.

Reviewing water point records

Water point records in the form of borehole drilling logs and pumping test reports from previous groundwater development projects are by far the most valuable source of information for the hydrogeological investigation. Such records are generally held by government departments responsible for water development, by donors and non-governmental organizations (NGOs) involved in rural water supply, and by conscientious drilling contractors. The amount of this type of information available will vary from one country to another and within each country. Often it will be necessary to supplement

---

Box 19: Problems at existing water points of inappropriate design

A review of rest-water level, yield and pumping test data for water points in six districts of Zambia's Southern Province shows that about one quarter (195) of all recently drilled boreholes, although sited in shallow groundwater, are low-yielding due to low aquifer permeability. These boreholes are of inappropriate design for the local groundwater conditions.

This conclusion is confirmed unanimously in discussion with local communities. At Manzambani Village, for example, 50 households have to wait for water at their new borehole and report that pumping the water is very hard. The shallow water table is quickly drawn down due to the low aquifer permeability and soon falls below pump level, causing people to have to wait for the water level to recover. In an attempt to overcome the problem, additional pipes were added (at additional expense) to place the pump deeper. This did not address the problem of low aquifer permeability and did not improve the situation, it just made it harder to pump water from an even greater depth.

Manzambani is a classic example of inappropriate use of a borehole where a large diameter well or collector well would be so much better, if only because of the increased storage provided as the well fills overnight. The situation was summed up by Headman Manzambani, when he said: "We could see the water, it was shallow, the water was coming in from the sides, but the driller went deep. Now the water goes deep because the pipes are deep. We don't know why he went deep".

A look at the principal rock types associated with this problem suggests that large diameter wells and collector wells will play an important role. A conservative estimate is that an area of 100 ha could be irrigated immediately by converting these boreholes to productive water points of appropriate design.

---

existing information with field measurements. However, a list of recent projects is often a good way to begin to search for where information is held.

Identifying water points poorly sited or of inappropriate design

From Figure 4.1, an existing borehole or well should potentially be resited where the recorded yield is low (say <0.3 l s⁻¹), the rest-water level is deep (say >15 m), drawdown on pumping is rapid, and recovery of the water level is slow. An existing borehole or well is of inappropriate design, and should potentially be converted to a screened regolith borehole, large-diameter well or collector well, where the recorded yield is low (<0.3 l s⁻¹), the rest-water level is shallow (<15 m), drawdown on pumping is rapid, and recovery of the water level is slow. Practical examples of each type of response during a pumping test are given below in Figure 4.3.

Identifying underutilized water points

Figure 4.2 shows how to identify underutilized water points based on a review of water point records and consultations with the local communities. Potentially underutilized water points will typically have recorded yields greater than 0.6 l s⁻¹ (ideally greater than 1.2 l s⁻¹) where pumping tests have been conducted, and ideally will have relatively shallow groundwater to facilitate ease of pumping.

A short-list of potentially underutilized water points drawn from the records can best be verified by the respective communities who will know whether the water points are reliable, particularly during drought. The first site visit can confirm the exact location of the water point, owner's name, date of construction, depth, diameter, rest-water level and pump type. By talking to the local community an estimate can be made of the local community structure and number of users, the average abstraction, the effect on water level and the rate of recovery, the water supply situation during recent drought years, and the potential demand for a productive water point. Box 20 provides an example of this rapid rural appraisal. If preliminary criteria are satisfied, the next step is to pump-test the water point to quantify the safe yield that will be sustained during drought.

Projecting safe yields during drought

The safe, sustainable yield of a borehole or well is uppermost in determining the productive potential. It can be reliably assessed only by carrying out a pumping test followed by projection of performance during drought and monitoring over the longer-term to confirm this performance and revise abstraction rates accordingly. This section provides a guide, giving practical examples and basic information to encourage the reader to take on this step which is key to both identifying underutilized water points and to siting new water points of appropriate design (Chapter 5).

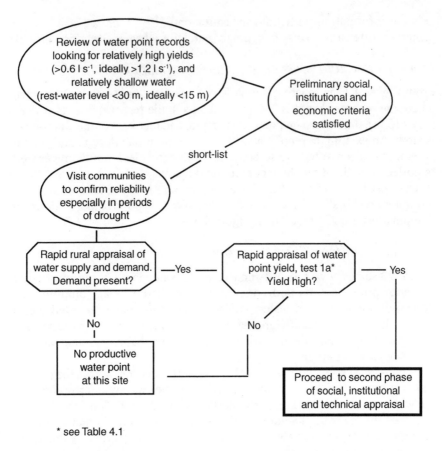

Figure 4.2 Identification of potentially underutilized water points

Types of pumping test

A pumping test simply involves drawing down the water level in the borehole or well by pumping at a known rate and recording the water level recovery with time. An example is the stepped-drawdown test. Pumping rate is increased in up to five roughly equal rates, while the water level in the pumped well or borehole is being measured. The pumping rate is then plotted against drawdown in water level to ascertain the maximum rate at which the water point can be pumped (Brassington, 1993).

However, the shallow depth and relatively low yields of many boreholes and wells in dryland areas mean that a stepped-drawdown test is often neither practical nor appropriate. In such cases, the maximum yield can be assessed by a more simple test where the water level is drawn to just above the pump inlet and then the pumping rate reduced to balance the inflow to the water point. This final pumping rate approximates the maximum yield that can be

Box 20: An example of rapid rural appraisal of a potentially underutilized water point

District (Communal land)	Zaka (Ndanga)
Ward (Chief)	34, Dzoro North (Bota)
Ag. Extension Worker (AEW)	Mr Makunde, Chivamba B/C
Place name (water point)	Muzondidya (borehole at Primary School)
Kraal(s) or village(s) served	Muzondidya, Gachiti, Maranele, Semende, Manyetu
VIDCO (Chairman)	1, Taivamunhamo (Mr V Semende)
Grid Reference	Map 2031C2; 366 176; 31°26'E 20°38'S Alt.620m
Initial Site ID by	AEW (Mr Maireva at that time)
Site Location	1km from main tar road Chiredzi – Zaka
Landform/Catchment	Large vlei, wide, low slopes, open land, at junction of small side vlei. RWL 5 m.
Geology	Granulite gneiss; area devoid of faults
Community Structure	5 kraals around proposed site. Muzondidya Business Centre 2 km.
Present Water: Domestic	Not critical: good B/hole at school provides whole community. Spring nearby until 1989. Two NORAD wells at B/C, and one on vlei give some water.
Production	Serious shortage. Spring 2km from potential site provides water for 10 individual gardens. Private land all around. Owner unwilling for community garden.
Previous Gardens	Co-operative garden for 30 families at Primary School, well now dry at 5m. 40 families at Chamuwonde had garden, 3 beds/family for home consumption.
Present Vegetables	Few from existing gardens.
Community Enthusiasm	Very good. Initial visit attended by many people including 3 kraal heads. Community proposed garden of 100 families, community well RWL 4m but inadequate. These kraals have co-operated previously. Would select members equally among 5 kraals wherever garden is sited.
To serve	250 families (domestic and gardening).
Land Allocation	Mr Chauke has offered land for the proposed community garden. Mr Marenele and Mr Gachiti would do same. Compensation required.
Market Opportunities	Appear excellent, main tar road close by.
Other factors	Much interest from menfolk of this community. Good spirit shown, active Master Farmers Club and VIDCO. Community Worker is Mrs J Musikavanhu.
Overall Impression	Distinct vlei landform, public, much interest, scheme primarily for gardening, excellent marketing opportunities. Chiredzi only 50 km.
Recommendation	Rapid appraisal of existing borehole at Primary School.

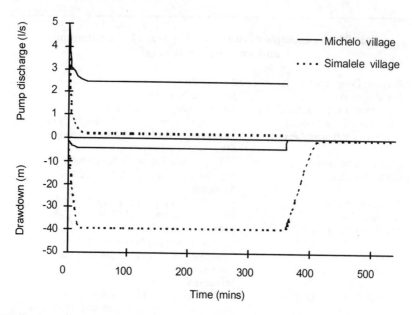

Figure 4.3 A simple six-hour pumping test on two contrasting water points in Zambia. The pump discharge rate was reduced to match inflow to the water point and drawdown remained constant until pumping stopped, when water level recovery was recorded. Michelo typifies an underutilized borehole (rest-water level 14.9 m, high discharge 2.5 l s^{-1} low drawdown 4.5 m, rapid recovery five minutes). Simalele typifies an inappropriate choice of well design (rest-water level 2.3 m, depth of borehole 57 m, low discharge 0.25 l s^{-1}, high drawdown 39 m, slow recovery 60 minutes)

maintained at that point in time. Figure 4.3 shows examples recorded by a drilling contractor in Zambia. They illustrate the contrasting water level responses to look for where a) the water point is strong and could support production if pump capacity were better matched to potential yield; and b) the water point is of inappropriate design in the local groundwater conditions but could support production if converted to a screened-regolith borehole, large-diameter well or collector well.

The maximum yield measured at a point in time by this method will vary as fluctuations in water table during the year and during drought change the available drawdown. It is prudent therefore to conduct more rigorous appraisal during development of productive water points, since this forms the basis for all subsequent planning of productive activities. If the water point and aquifer stand up to more rigorous appraisal, the long-term reliability can be anticipated with greater confidence. The following is a description of the procedure used successfully in southern Africa. It is based on a set of constant-rate pumping tests, listed in Table 4.1, and modelling of recorded water level response to project safe yield during drought.

Timing of the appraisal

The tests listed in Table 4.1 should be undertaken in the dry season wherever possible, when peak water demand occurs in practice and when the yield and water-table are at their lowest in minimum depths of saturated aquifer. The ideal time is actually at the end of an extended dry cycle (Figure 2.5). Water points found to be reliable in these low groundwater conditions will withstand future periods of drought, but no similar guarantee can be made for water points tested during high groundwater conditions.

Table 4.1 Pumping tests performed during appraisal of potentially underutilized water points and during siting and selection of appropriate new well designs (T = transmissivity; S = storativity)

Test	Description	Pump time (min)	Pump rate (l s⁻¹)	Test Results
1a	Rapid appraisal of existing borehole using handpump fitted	60	0.5	T, S, yield
1b	Appraisal of existing borehole and new screened regolith borehole using electric submersible pump	300	0.5	T, S, yield
1c	Appraisal of 15–20 m exploratory hole	60	0.2	T, S, projected LDW yield
2	Appraisal of existing family well	60	1.0	T, S, yield, projected LDW yield
3	Appraisal of new large-diameter well (LDW)	300	1.0	T, S, yield
4	Appraisal of new collector well	300	1.0	T, S, yield, % improvement by radial drilling

Equipment

Brassington (1993) illustrates equipment commonly used in pumping tests. The following have been found helpful to perform the tests listed in Table 4.1:

○ electric submersible pump rated at 0.8 l s⁻¹ at 40 m head with starting capacitor and 50 m of waterproof electric cable and connectors
○ 1.5 kW petrol generator to power the electric submersible pump
○ 70 m × 50 mm lay-flat discharge hose with quick release connectors
○ discharge hose elevation unit
○ 50 mm gate valve for flow-rate control and appropriate pipe fittings
○ orifice plate flow meter with 19 and 30.6 mm orifice plates
○ 40 mm water meter for measuring pumped volume (e.g. Kent PSM)
○ three manual, battery-operated water-level dippers
○ automatic water-level recorder e.g. Munro chart recorder

- two stopwatches, clipboards and pens
- lap-top computer and printer
- computer power unit (e.g. 12v dc to 240v AC converter) and car battery to allow field data entry
- camping equipment for field staff where necessary
- tools for removing handpumps e.g. SIWIL pipe lifter
- electrical conductivity meter to measure groundwater salinity.

Flow measurement

Figure 4.4 illustrates the equipment layout. To perform an accurate test the pumping or flow-rate should remain constant. A positive displacement pump (e.g. Mono-pump) can provide a constant rate due to the relatively flat head-to-flow-rate characteristic of this design, but these pumps are awkward to transport and install and require bulky power units. An electric submersible pump, on the other hand, is easier to use and thus suited to rapid testing, but has a steeper head-to-flow-rate pump characteristic. This means that as the water level in the well or borehole falls during the test, the increased pumping head causes a reduction in the pumping rate. To counteract this and to keep the pumping rate constant, a gate valve on the discharge pipe is opened to maintain manometer water levels. These continuously measure pressure difference across the orifice plate and reflect the flow rate.

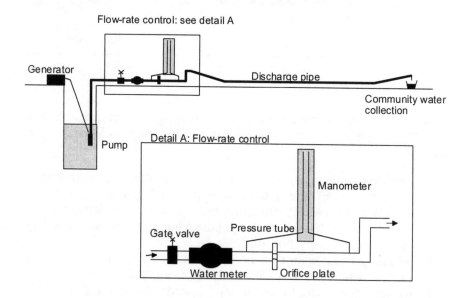

Figure 4.4 Layout of pumping test equipment

The pressure difference / flow rate characteristics of 19 mm and 30.6 mm orifice plates are illustrated in Figure 4.5. The water meter is used to measure total pumped volume. It should run full by ensuring that the outlet is higher than the inlet. It also allows flow rate to be checked occasionally by timing the passage of 10 litres, as a backup to the orifice plate flow-meter. When testing in dry areas, local people will want to use the pumped water. If the discharge pipe is lying on the ground they will lift it to fill their buckets and the varying back pressure will cause inaccurate flow-rate measurement. To prevent this, the end of the discharge pipe should be elevated 0.5 m in a smooth arc to allow ready access for their buckets.

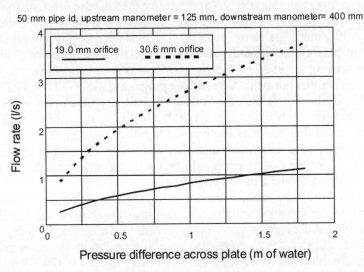

50 mm pipe id, upstream manometer = 125 mm, downstream manometer= 400 mm

Figure 4.5 Rating curves for orifice plate flow-meter

Personnel

Successful testing requires careful data collection and analysis. Ideally this should be the responsibility of one person who manages the field activities, analyses the data and makes the decisions based on the results. This field technician is assisted by local members of the community, who are responsible for handpump removal and replacement and long term water-level monitoring before and after the test. They will reside on site for the duration (typically 24 to 96 hours) allowing the technician to supervise only the pumping and be able to organize several tests at different sites simultaneously.

Water-level monitoring

Manual battery-powered water-level dippers are adequate for monitoring water level during short tests. The water level should be monitored from the time local people stop abstracting water prior to the test through to the complete recovery of the water level after the test. A water-level chart recorder is a useful way to measure water level continuously, again reducing the time that the field technician must be on site. Dipping frequency during tests will vary from every half a minute at the start of pumping and the start of water level recovery, to every six hours towards the end of a test.

Analysis of the pumping test data

Pumping test data can be analysed either manually or with the aid of a computer and proprietary software package. In manual analysis, the data are plotted graphically and curve fitting is used to determine values for aquifer transmissivity (T) and storativity (S) which are then used to estimate the safe maximum discharge for a given available drawdown in the well or borehole at any given time. The same steps of graphical representation and curve fitting are performed automatically by computer, and values of T and S obtained are again used to estimate safe maximum discharge. Where computing facilities are available, software programs BGS PTFIT and PTSIM, developed by Barker (1989) for both small and large diameter wells, are now freely available on the British Geological Survey website (www.bgs.ac.uk/bgspt). They can be used to estimate values of T and S and to model water level behaviour under a given abstraction regime for a set period of no groundwater recharge. In this way, the safe yield that will be sustained during drought can be projected.

Appendix 2 provides practical examples of pumping test data and analysis. The example is for a potentially underutilized borehole at Muzondidya Primary School. Review of water point records and a first community visit (Box 20) found that the conventional borehole drilled by the state had an approximate 'blowing' yield at the time of drilling of 2.1 l s⁻¹ (50–75 m³ d⁻¹). However, abstraction by the local users was only 5–10 m³ d⁻¹ using the single handpump fitted. Rapid appraisal by test 1a (Table 4.1) using the handpump fitted, and manual interpretation of the data using the simple straight line Cooper and Jacob technique, shows the estimated safe maximum yield to be of the order 100 m³ d⁻¹. Indeed, the water level in the borehole dropped only 3 m during pumping and recovered very quickly, a good indication of a high-yielding borehole.

More rigorous analysis using program BGS PTFIT shows estimated aquifer properties T=32 m² d⁻¹ and S=2 × 10⁻⁶. An excellent fit between recorded test data and modelled data provides confidence in the result. Using these aquifer properties in program PTSIM allows the safe yield to be projected over an extended dry period set at 240 days. This confirms that the borehole is underutilized, having a potential safe yield of the order of 200 m³ d⁻¹ or 20 times greater than the present pump capacity.

It must be stressed that equations used in the analysis of pumping test data are based on many assumptions, most of which are not met in reality in hard rock aquifers, and the results should be viewed only as 'order of magnitude' values and not precise predictions. They do, however, provide a useful basis for planning, as long as they are used in conjunction with long-term monitoring of the groundwater resource and any environmental impact, with abstraction rates being revised accordingly.

Equating safe yield to irrigable area

On determining the safe yield, a community whose priority is to use the water point to support small-scale irrigation will want to know the area of land that can safely be irrigated. There are two approaches to answering this question: theoretical and practical.

Theoretical calculation of irrigable area

In theory, the wetted area that can safely be irrigated at a productive water point can be calculated by the following formula:

$$A = [Y - (N_P \times C_P / 1000) - (N_L \times C_L / 1000)] / (E_t \times K / 1000)$$

where A = wetted area (m²)
Y = water point sustainable yield (m³ d⁻¹)
N_P = number of people
C_P = consumption level per person (l d⁻¹)
N_L = number of livestock
C_L = consumption level per livestock (l d⁻¹)
E_t = maximum potential evaporation rate (mm d⁻¹)
K = peak crop factor

Experience shows that people in dryland areas with water points near to their houses will often use 20 to 40 litres of water per person per day (lpd), decreasing to consumption levels of less than 10 lpd for people fetching water from more distant sources. Figures for livestock consumption rates range from 20–40 litres per head per day (lhd) for cattle down to 1–5 lhd for sheep and goats. Maximum potential evaporation rates recorded in dryland areas are typically in the range 4 to 8 mm d⁻¹, and the maximum crop factor used to determine the peak irrigation water requirement of a mature, relatively thirsty crop such as maize is typically 1.2.

Design example

Data: - Water point with sustainable yield of 30 m³ d⁻¹ (Y)
- Community of 130 families, average family size 7, number of people 910 (N_P)
- Average consumption level per person 20 lpd (C_P)
- No. of cattle 30, no. of goats 50, n. of livestock 80 (N_L)

- Average consumption level per livestock 13 lpd (C_L)
- Maximum potential evaporation rate 6 mm d^{-1} (E_t)
- Peak crop factor 1.2 (K)

Wetted area = [30 - (910 x 0.020) - (80 x 0.013)] / (6 x 0.0012)
         = 1500 m^2 or 0.15 ha

For allotment-type gardens comprising individual family plots surrounded by pathways, the fenced area needed will generally be 1.6 times greater than the wetted or irrigated area, and in this example would correspond to a fenced garden of total area 0.24 hectares.

Practical calculation of irrigable area

In practice, it has been found that the area that people choose to irrigate is larger, and the volume of water taken for domestic use is less, than the theoretical projections. Figure 2.3 shows monitoring data recorded during and after the serious drought of 1991-92 at a productive water point in Zimbabwe. The scheme at Romwe, with peak water abstractions of up to 100 m^3 a week, is serving part of the domestic needs of 103 families and irrigating a fenced community garden of 0.4 ha. The partition of water use to irrigation and domestic supply is shown further in Table 4.2.

Table 4.2 Partition of water use to irrigation and domestic supply at seven productive water points

	Romwe	Muzondidya	Gokota	Dekeza	Nemauka	Mawadze	Matedze	Average
Safe yield (m^3 d^{-1})	10.8	18.0	17.1	34.1	40.2	34.7	18.3	24.7
Average abstraction (m^3 d^{-1})	6.6	5.2	7.7	10.3	4.3	6.6	9.6	7.3
Peak abstraction (m^3 d^{-1})	15.1	21.3	18.9	22.1	22.8	20.3	22.8	20.5
Families taking domestic	103	21	93	60	116	89	110	84.6
No. of people taking domestic	721	171	711	436	874	771	919	658
Domestic use: all sources (m^3 d^{-1})	10.1	2.3	10.0	6.0	11.6	9.1	11.5	8.7
Domestic use per person (lpd)	14.0	13.4	14.1	13.8	13.3	11.8	12.5	13.3
Domestic from water point (lpd)	1.7	7.0	5.2	6.4	1.0	3.1	2.4	3.2
Domestic as % of abstraction	19	20	49	27	22	36	23	28
Wetted area in garden (ha)	0.25	0.31	0.31	0.31	0.31	0.31	0.31	0.3
Average irrigation applied (mm d^{-1})	2.2	1.7	1.3	2.5	1.1	1.4	2.5	1.8
Peak irrigation applied (mm d^{-1})	5.1	5.7	3.2	5.4	5.9	4.3	5.9	5.1

On average, people at these productive water points use 13 litres of domestic water per person per day, of which about one quarter is drawn from the productive water points and three quarters from other sources, which vary depending on the time of year. Livestock numbers are negligible due to massive losses during the drought of 1991/92. On average, 72 per cent of water (1900 m^3 yr^{-1}) drawn from the productive water points is used to irrigate the gardens. Considering the peak irrigation requirements recorded (which determine the maximum area that can be irrigated) and the proportion of water typically drawn for domestic purposes, a general rule of thumb to help determine the irrigable area at a productive water point is:

Wetted area (ha) = 0.02 × safe yield of the water point (m^3 d^{-1})
and
Fenced area (ha) = 1.6 x wetted area (ha)

Using this general rule of thumb, the irrigable area determined from field evidence is typically four times greater than the theoretical calculation, highlighting that, in practice, smallholders are able to make effective use of limited water resources and still achieve worthwhile production. When planning, it is important that estimates of crop water requirements are in harmony with farmers' practices and requirements. Over-conservativeness leads to over-design, which can result in viable projects being turned down on the grounds of inadequate water resources or excessive cost. Planners and engineers should be positioned to take the lower water requirements of smallholders into account. It is also important to consider the area of land that is fenced from a social perspective. Although it may appear sensible to start small (perhaps using the theoretical calculation of irrigation) and expand the garden based on monitoring of water point performance, in practice the physical presence of a fence around a relatively small, initial area can prove to be an important constraint to membership and equity, which in the long-term are better served by fencing a larger area at the outset, based on projected safe yield of the water point, and allowing the community to manage their own water supply accordingly. Aspects of equity and membership are considered further in Chapters 6 and 7.

Choosing the most appropriate pump technology

Although it occasionally does occur that artesian water will rise above the ground surface, doing away with the need for pumping, in most cases it is necessary to bring well or borehole water to the surface with the help of a pump. The subject of appropriate pump technology is vast, so this section will cover only technical criteria important to a productive water point. The social factors of community ownership and incentive that are key to sustainability, and the steps taken during scheme implementation to build capacity for village-level operation and maintenance are detailed in Chapter 6.

The choice of pump technology at a productive water point is critical because the pump(s) will lie at the heart of the production system. Whether to support small-scale irrigation or livestock or other productive activity, inappropriate choice of pump technology will lead to ineffective use of the productive water point and jeopardise long-term sustainability of the scheme. There is no single design that is ideal for all situations, but the features listed in Box 21 may help when choosing pump technology for a productive water point. Of these features, the robustness of the design chosen has been found to be particularly important (Box 22). It is also important to know if the water has a corrosive nature, as this will require the use of special materials in pumps, pipes, well screens, etc. The main characteristics of corrosive water are:

Box 21: Checklist when choosing pump technology

O The pump(s) should enable the capacity of the water point to be fully and correctly exploited, drawing water at a rate up to the projected safe yield from the full depth of the water point – only in this way can the benefits of investment in a relatively high-yielding water point be fully realized.

O However, productive water points should be structured for independent management by the users (as opposed to requiring management by external agencies). The choice of technology should depend on what is manageable for the users, instead of the technology dictating the management structure. It should reflect their wishes but must be appropriate to the setting, striking a balance between the socio-economic conditions and the community's aspirations and abilities.

O Sufficient thought is needed at the planning stage to the O&M arrangements that will be required, to who will undertake repairs, to the way running costs will be met, and to the expected life of the pump(s); the pumps used must be such that the users can afford to maintain them.

O Durability of the lifting device is of prime importance, particularly where the pump(s) will be handled by many people rather than a few individuals.

O Parts in contact with the water must not corrode, contaminate or give a foul taste. Any materials that rust, are subject to rapid abrasion, or which are toxic (including some PVC pipes) should be banned.

O The pumps should use affordable and readily available replacement parts; cost of replacement rather than user-friendliness is often a key factor in consumer preference.

O The pumps should be insensitive to the quality of drilling. Local drillers may not always deliver vertical boreholes. The entire system must be capable of being pulled from the water point as required.

O Repair facilities must be available – it is sometimes more economical to have simple, less efficient equipment that can be repaired quickly and easily on the spot using readily available tools and without need for heavy lifting gear, rather than more efficient devices liable to serious mechanical failure that take longer and cost more to repair.

O Multiple pumps should be fitted to the same water point if this allows appropriate user-friendly technology to be used and matches the required pump capacity; multiple pumps also provide versatility as the population increases or as per capita water requirements expand, and they guarantee some abstraction when one pump breaks down.

O Technology options should also include phased development that includes training of users in the management, operation and maintenance of progressively more effective mechanized lifting devices.

O The relative costs of running different types of mechanized pumps should be fully considered. Potential contamination of the water source, and the disadvantage that mechanized pumps can run dry and cause damage, should not be overlooked.

O Meteorological data should be checked carefully to confirm viability where wind or solar-powered pumping devices are being considered; they are not technically viable in all dryland areas.

pH <7; total dissolved solids >1000 mg l⁻¹; chloride (Cl) >500 mg l⁻¹; dissolved oxygen (O_2) >2 mg l⁻¹; hydrogen sulphide (H_2S) >1 mg l⁻¹; free carbon dioxide (CO_2) >50 mg l⁻¹. Groundwater quality is discussed further in Chapter 5.

Manual pumps for productive water points

Water lifting devices fall into various categories: handpumps, footpumps, animal-driven pumps, mechanically-driven pumps, suction pumps to 6 m head, medium-lift pumps (7– 20 m head) and deep-lift pumps (>20 m head). Table 4.3 lists some of the available technologies that will be suitable for productive water points in different settings. Cost estimates are for complete units to the depth (or head) shown, although prices will vary from country to country depending on local manufacture, import duties, exchange rates, etc.

Manual lifting is limited by the physical power of the lifter, and decreases with increasing depth from which water is being raised. Typical abstraction rates by one person using a handpump are relative low. Footpumps, which use the powerful leg muscles more effectively, allow pumping at higher outputs for longer periods. Hence, treadle pumps will be important where groundwater is shallow.

Where groundwater is deeper, the area that can be irrigated using a manual pump decreases, and at depths greater than 20 m only a small area of land (about 0.03 ha) can be irrigated by one person. In such cases, multiple footpumps (such as the Vergnet hydraulic diaphragm pump) fitted to the same water point, or a rotary-drive handpump with flywheel and large-diameter, stainless steel cylinder to increase capacity and reduce friction losses, will be important to make effective use of a relatively high-yielding borehole.

Mechanized pumps for productive water points .

Where productive water points support irrigation of less than about one hectare, and where groundwater is relatively shallow, several manual pumps fitted to the water point will be a sensible option. However, where the safe yield of the water point is relatively high (>40-50 m³ day⁻¹) and or where groundwater is relatively deep (>30 m), mechanized pumps will be needed to make effective use of the water resource (Figure 4.6).

The use of mechanized pumps at public water points is contentious, plagued by social and technical problems in past domestic water supply programmes. Although it is important to recognize this experience, it is also important to recognize that social, technical and economic appraisal of productive water points (Chapters 1 to 3) clearly shows that mechanized pumps will be increasingly important to communities wishing to maximize benefits in terms of production, membership, equity, time-saving, benefit-to-cost ratio and internal rates of financial return.

In many cases, human-powered devices offer the best means to initiate small-scale irrigation because of their low first cost and less demanding

Table 4.3   Some water lifting devices suitable for productive water points
(adapted from Fraenkel, 1997)

	Example	Capacity (m³ hr⁻¹)	Cost (US$)	Comment
Surface water 0–3 m head				
Swing basket		5–7		
Pivoting gutter	Dhone	5–10		
Diaphragm pump	Bumi, IRRI	2–20	15–200	Ideal for shallow wells and dams
Archimedean screw		15–30		
Mechanized scoop wheel	Sakia, Tabla	36–360		Motor powered
Propellor (axial flow) pump	IRRI	100–500		5 hp engine or 3 kW motor
Shallow water 0–7 m head				
Rower pump	Nsimbi, TARA, Maida	1–3	50–70	
Counterpoise lift	Shadoof	2–4	10–20	
Coil or spiral pumps		2–10	50–70	
Self-emptying bucket	Mohte	5–15	10–20	Animal powered
Persian Wheel	Zawaffa, Jhallar, Noria	5–150		Animal powered
Hydraulic handpump	H20 E-zee flow	2–3	250	Can lift to tank 20 m high.
Treadle or pedal pump	ApproTec Moneymaker	5–7	70–150	Ideal for shallow wells/dams.
Pressurised treadle pump		1–2	250–500	Can lift to tank 20 m high.
Low-lift centrifugal pump	2–5 HP	5–7	700–1000	Petrol driven
Medium water 0–30 m				
Bucket and windlass		0.5–1	10–20	Multiple units on wide wells
Chain (or rope) and washer	Chinese liberation	1–3	70–180	
Hand lever piston pump (75 mm diameter cylinder)	Afridev, India Mk II	1–2	500–800	Suitable in numbers; larger cylinders may be an option.
Bicycle-driven mono pump	Monolift DW90	1–2	1000	Reputation for reliability
Hydraulic action footpump	Vergnet 4C Hydropump	1–2	700	3 units fit on ID 150 mm, 4 on ID 175 mm. Low annual costs (US$10–30), easy repair from surface, corrosion-resistant parts.
Motorized chain and washer		20–30	1000	Electric motor powered
Self-priming jet centrifugal combination pump	1.5 HP	2–3	500–1000	Useful for lifting dirty water, or where source may sometimes be pumped dry.
Spring rebound inertia pump (or water oscillation pump)	Village Life	1	1745	3 pump unit (Trident Triple, US$5235, yield 2.7 m³/hr at 30m) fits on ID 100 mm. Combined hand and foot lever, low annual costs, corrosion resistant parts
Deep water 0–100 m				
Flywheel drive handpump	Volanta 2000, Nira	0.5–2	2000	Low annual cost (US$10–20)
Hydraulic action footpump	Vergnet 4C Hydropump	0.5–1	800	As Hydropump above.
Multi-stage centrifugal pump	Electric submersible	1–100	3000	With 1.5kW petrol generator
Conventional centrifugal pump	0.03–16 kW	5–200	3000	Most common motor pump for small scale irrigation.
Progressive cavity	BH155 Mono pump and drive belt to Lister LT1 diesel enginer	5–200	3000	Reputation for reliability, fits down slim boreholes, copes better with water level variations than centrifugal pump.
Hot-air pumping engine	Rider-Ericsson 0.5 HP (Stirling) c.1900	2–3		Can use any combustible fuel. May not still be made.
IT windpump	Kijito, Tawana	3 m s⁻¹ wind; 15 m³ hr⁻¹ at 40 m head 8 m³ hr⁻¹ at 80 m	7000	Cost-effective where mean windspeeds greater than 4 m s⁻¹, not where less than 2.5 m s⁻¹. but are not cost-competitive where windspeeds are signiicantly below 2.5 m s⁻¹
Solar	Mono Solarlift SW4L	2–3 m³ hr⁻¹ at 40 m head	15000	High iniital capital cost but low recurrent cost. Theft of panels can be a problem.

Box 22: Use robust pumps on productive water points

Most handpumps have been developed for use by a family to provide mainly drinking water for themselves and their livestock, rather than for irrigation. The problem when pumps of this kind are used for irrigation is the intensity with which they are used; instead of pumping for a few minutes per day they have to be used for several hours per day, which naturally tends to shorten their useful lives and to increase the incidence of breakage. The graphs show the incidence of breakage of Zimbabwe 'B' type handpumps across nine productive water points over a period of five years. Each water point has two pumps drawing water from up to 18 m depth.

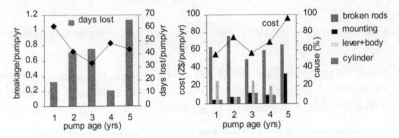

Although considered robust under domestic conditions and accepted as the standard in Zimbabwe, these lever-type pumps have faired badly in communal small-scale irrigation. Weak points are the fulcrum mechanism and pump body which have cracked through metal fatigue; no bracing to the pump bodies which allows vibration in the concrete mounting; breakage in the drive train due to the operator 'swinging' on the handle or hammering the end stops, the pump rods going 'slack' as the hand lever is lifted too rapidly before taking up the load again on the down stroke.

Rotary-drive handpumps do not suffer from the damage common with lever pumps. Here, the piston is driven by a crank from a flywheel which smoothes the fluctuations and makes the pump easier and more efficient to work, especially for long periods. The cyclic loading involved in accelerating discrete cylinder-volumes of water up the rising main is absorbed by the flywheel's momentum and not felt by the operator. The disadvantage is that rotary drive pumps cost more initially, and are heavy due to the flywheel and crank mechanism and requisite supporting column. But for pumping large quantities of water for irrigation, and for pumping by hand from depth, lower annual costs for repair and the improved ergonomics make the flywheel adaptation sensible, especially where below-the-ground components can be identical to those of lever pumps already manufactured in-country.

Higher initial capital cost is a feature common to all types of good quality pump, but is justified by the reduced frequency of the expensive operation of removing and overhauling the pump. Moreover, most groundwaters are corrosive to some degree, and only quality pumps using corrosion-resistant components will have a life expectancy of 10–20 years. The result of using alternative 'cheap' pumps is manifest in the number of boreholes that are lost to future use when the pump finally corrodes to the extent that a part falls off when being extracted for repair. Not only has the pump to be replaced but also the borehole has to be redrilled – all because of a minor initial capital cost saving.

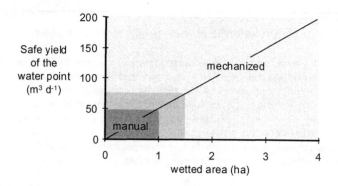

Figure 4.6 Water point yield and irrigable area, and the corresponding roles for manual and mechanized pump technologies

institutional arrangements for O&M. However, in the longer term, it is hoped that communities will be assisted to advance towards more productive pumping techniques, which inevitably require some mechanization (Box 23).

Box 23: Manual or mechanized pumps?

'Contrary to popular belief, human muscular energy is not cheap. The poor are forced to use human power, usually because they cannot afford anything better, since the cash investment required is minimized and therefore is more "affordable" than other options. As will be shown, almost any other source of power will pump water more cheaply unless only very small quantities are required.

The human work capability is around 250 Wh/day, so it takes four days of hard labour to deliver just 1 kWh; this is the output which a small engine could deliver in less than one hour while burning less than one litre of petroleum fuel. So the farmer with a small mechanized pumping system has the equivalent of a gang of 20 to 40 people who will work for a "wage" or running cost equivalent to, say, 1 litre of fuel per hour. Not surprisingly, any farmer who can afford it will choose to employ an engine rather than 20 to 40 people. This argument can be turned on its head to show the high price of human muscle power; if the "opportunity cost" or a real wage cost is assigned to human muscular labour; e.g. assuming a daily wage rate of US$1.00/day gives an energy cost of about $4.00/kWh. Although this is a low wage for hard labour, even in some of the poorer countries, it represents an energy cost that is significantly more expensive than even new and exotic power sources such as solar photo-voltaic panels.

There is an opportunity cost caused by diverting people from more important work to pumping water; the best asset people have is brains rather than muscle; therefore, if agricultural productivity is to improve and economic standards are to be advanced, it is essential to introduce more productive power sources for all except the very smallest of land-holdings.'

Source: Fraenkel (1997)

Box 24: Who chooses the hardware ?

Many schemes aiming to supply water to rural and peri-urban communities target the lowest capital cost of equipment to be installed. Even if the project implementer is experienced, the tendency is to limit the scope of supply contracts to the most basic elements in the belief that for a large number of pumps one can negotiate the 'best' deal by buying in one tender all the units likely to be required. The implementer chooses the 'hardware'.

To some, this flies in the face of good buying practice. A 1994 UNDP-World Bank Water and Sanitation Report comments: 'It appears to have been accepted that the price range for purchase of a VLOM handpump should be between US$500-1000. Field experience has now shown that this is not a good economic guideline to follow, but rather that the maintenance cost and lifespan of the handpump be considered. Thus a good handpump is also one which the community can afford to look after. Value for money must not be judged on purchase price alone, which has been a common mistake in the past.'

From an economic perspective, installation cost, maintenance cost (including wearing parts and labour to replace), user training cost, performance (including efficiency and output) and equipment life must also be considered. Why not seek a 20 to 25 year life over which all these expenses can be amortized?

From a social perspective, another factor which should be evident for any scheme which seeks to hand responsibility for sustainability to the users is the choice of the users themselves – it is a question about which more financiers and implementers are at least thinking today as they are asked to rehabilitate schemes less than a decade old. Including the users in the 'hardware' choice might at first sight seem a recipe for clouding the issue but if they are expected to foot a part of the capital cost (in cash or kind), as is now the priority, then it is socially irresponsible to not involve them. In fact, it is the users who need to be informed about annual costs – if unable to afford capital costs it is unlikely they will be able to afford 'high' annual costs. To encumber a community in this way is to redefine a sustainable project as one in which the users cannot afford maintenance without external (implementer) support. We should not underestimate the ability of users to recognize such a millstone and not be surprised when they 'opt out'.

Consideration by the users at the outset will help to avoid frequent future complaints over low delivery rate, reliability and queuing at water points. If there is a comparison on site, they will quickly identify the reliable, easy-to-use pumps with higher volume capability that remove these issues and open the way to their taking ownership and long-term responsibility. Rather than the arms length approach adopted by some implementers, field experience suggests that the manufacturer or supplier of the 'hardware' should always be brought into the equation of creating a project from the start, to ensure that communities are aware of the range and merits of different pump technologiese, and that the problems associated with appropriate materials selection, spare parts distribution and maintenance training are also a part of the supply contract.

Provided by John Carter, an independent Pump Adviser with over 20 years experience in rural and peri-urban pumping systems and who works with hydraulic action human powered pumps in East, Central & Southern Africa. He can be contacted by e-mail at pslafrica@aol.com.

It is vital that prospective participants at productive water points have full participation in the planning process and in the choice of technology, as they will eventually be using the infrastructure on a daily basis and will be faced with the task of making it succeed despite its inevitable limitations (Box 24). An important distinction between productive water points and conventional domestic supply is that, in common with all successful irrigation schemes, they are 'farmer-managed'. Being owned and operated by the community creates both the cash and the incentive needed to cope with the demands of mechanization. Hence, productive water points extend the range of appropriate pump technologies and mechanization becomes viable, *if* sufficient attention is paid to the local institutional arrangements required. This key investment in social development during scheme implementation is detailed in Chapters 6 and 7.

References and further reading

Alberts, H., Meza, R., Solis, D. and Rodriguez, M. (1993). How the rope pump won in Nicaragua. *Waterlines*, 12 (2) 3-5.

Arlosoroff, S. (1987). *Community Water Supply: the handpump option.* World Bank, Washington.

Barker, J. (1989). Programs to simulate and analyse pumping tests in large diameter wells. *Technical Report WD/89/24 (updated 1994),* British Geological Survey, Wallingford, UK.

Brassington, R. (1993). *Field Hydrogeology. Geological Society of London Professional Handbook.* John Wiley & Sons, Chichester, UK.

Chleq, J.L. and Dupriez, H. (1988). *Vanishing land and water: soil and water conservation in dry lands.* Land and Life Series. Macmillan Publishers, London, UK.

Clark, L. (1996). *The field guide to water wells and boreholes.* John Wiley and Sons, Chichester.

Doorenbos, J. and Pruitt, W.O. (1977). Crop water requirements. *FAO Irrigation and Drainage Paper No. 24,* FAO, Rome. 114pp.

Elson, R. J. and Shaw, R. J. (1993). Technical brief No. 35: low-lift irrigation pumps. *Waterlines* 11 (3), Intermediate Technology Publications, London.

Euroconsult (1989). *Agricultural compendium for rural development in the tropics and subtropics.* Elsevier. 740pp.

Fraenkel, P. (1997). *Water-Pumping Devices: a handbook for users and choosers.* Second Edition. Intermediate Technology Publications, London, UK.

Franceys, R. (1987). Technical brief No. 13: Handpumps. *Waterlines* 6(1) Intermediate Technology Publications, London.

IRC (1988). Handpumps: issues and concepts in rural water supply programmes. *Technical paper No. 25,* International Water and Sanitation Centre, The Hague, The Netherlands

Kruseman, G.P. and De Ridder, N.A. (1991) Analysis and evaluation of pumping test data. 2nd edition. *ILRI Publication 47.* Int. Inst. Land Reclamation and Improvement. Wageningen, PO Box 45, 6700, AA, The Netherlands.

Lambert, R. A. (1990). *How to Make a Rope and Washer Pump.* Intermediate Technology Publications, London, UK.

Lambert, R. A. (1992). The need for and design of a pressurised discharge human-powered treadle pump. *Paper 9839. Proc. Instn. Civ. Engrs Civ. Engng.* 92, 66-73.

Lambert, R. A. and Faulkner, R. D. (1989). A low-cost irrigation package for micro-scale irrigation in Africa. In: Dodd, V.A. and Grace, P.M. (eds.) *Proc. 11th International Congress on Agricultural Engineering,* Dublin.

Meier, P.M., Sanchez-vila, X. and Carrera, J. (1997). Study of transient constant rate pumping tests in heterogeneous media. *Proc. Rabat Symposium: Hard rock hydrosystems.* IAHS Publ. No. 241, 135-149.

Reynolds, J. (1992). Handpumps: toward a sustainable technology: research and development during the water supply and sanitation decade. *Water and sanitation report, UNDP World Bank Water and Sanitation Program,* World Bank, New York.

Sakthivadivel, R. and Rushton, K.R. (1989). Numerical analysis of large diameter wells with a seepage face. *J. Hydrology* 107, 43-55.

Thompson, D.M. and Lovell, C.J. (1996). Small scale irrigation using collector wells pilot project, Zimbabwe: Hydrogeological evaluation of potential areas by exploratory drilling and pumping test analysis. *ODA Report 95/16,* Institute of Hydrology, Wallingford, UK.

Thompson, D.M., Lovell, C.J., Chilton, P.J. and Macdonald, D.M.J., 1995, *Small scale irrigation using collector wells pilot project, Zimbabwe: Sites Report.* British Geological Survey Technical Report WC/95/75, Wallingford, UK.

Zanamwe, S. and Van Harderwijk, W. (1994). The SIWIL pipe-lifting device. *Waterlines* 12 (3) 30-31. Intermediate Technology Publications, London, UK.

5 Developing new water points of appropriate design

Introduction

AS WE HAVE SEEN IN Chapter 2, basement aquifers in dryland areas are complex in occurrence and their high spatial variability and low permeability make development of the groundwater resource relatively difficult. As a result, current practice is to dig traditional wells or drill deep boreholes as point sources for domestic supply, fitted with a handpump, bucket-pump or bucket and windlass. The total abstraction for a rural community by this degree of development amounts to the equivalent of about 0.5 mm of recharge, where present evidence is that actual recharge can be much higher, in the range 50–150 mm. There is therefore considerable potential which the present development methods are not capable of utilizing.

Dijon (1987) commented on the dilemma:

> In crystalline rock areas, 'appropriate technology' cannot in any way signify simple, locally engineered, or cheap technology. For a long time, African villagers have been trying to construct wells using simple tools and local materials. Such wells cannot be driven beyond the alteration zone, as what is required is a combination of jackhammers and explosives or the use of powerful drilling rigs to penetrate the hard rocks. Very little can be done to improve traditional methods, and the African people have little to learn from the benevolent organizations which come to help them with good-hearted volunteers, shovels and barrows. In the end, it is unavoidable to use imported, sophisticated and more costly technologies.

This is not to say that expensive deep boreholes are the only answer, or that shallow dug wells have no place: on the contrary, low-cost family wells are the first and logical choice in areas where they can readily provide sufficient reliable water. It is important to recognize, however, that new methods will, in many cases, be better suited to the local groundwater conditions, and that for many families living in dryland areas, these new 'appropriate technologies' will be necessary if a more equitable supply of water and the potential of the groundwater resource are to be realized.

Striking a balance between low-cost traditional means and new 'appropriate technologies' is therefore a very important process, which is best established through a systematic evaluation of all potential sources of water to meet present and future demand. Chapter 3 provides decision support from an economic perspective, it shows that a range of technology options are available which are economically viable. This chapter details steps taken to site and select the most appropriate (cost-effective) well designs, in areas where new source construction is required.

Finding water

In current practice, there are three main approaches to siting new water points, used either individually or in combination:

○ water divining by a water diviner or dowser, usually a local person who has a reputation for possessing the skill of being able to locate places to dig a well;
○ review of local groundwater conditions and natural indicators;
○ review of aerial photographs, aeromagnetic maps and satellite imagery to detect surface features such as faults and dykes, followed by a geophysical survey at selected locations, usually to identify whether fracture systems exist at depth.

Water divining and review of local groundwater conditions

So far, water divining has not been shown to work when subjected to objective scientific examination, and practical experience based on knowledge of the behaviour of other wells in the area is generally considered a better way of siting a water point. In India, for example, hydrogeological exploration to locate sites for hand-dug wells involves collecting detailed information from all existing wells in an area. An attempt is always made to understand why a certain well is yielding a good supply while another has failed. Box 25 lists some natural indicators and observations considered important from around the world. It should be remembered, however, that in hard rock aquifers, levels of abstraction sustained by wells sited using *only* this approach are generally low (Chapter 2) and rarely sufficient for production by all in a rural community who wish to participate.

Geophysical siting of boreholes

Deep boreholes (or tubewells) of diameter 0.1–0.2 m are now the most widely constructed type of water point in many dryland areas. Being mechanically drilled they are relatively easy to construct, and with modern air-driven rotary percussion can be drilled in a relatively short time. Few are completed in the regolith and most penetrate the bedrock in search of fractures which yield water. A depth of 60–70 m is typical in rural water supply. The bedrock section is generally left open but the shallow weathered zone is cased off

Box 25: Finding water – a checklist of natural indicators

○ the thickness of the weathered rock and whether it increases or decreases in topographical lows;

○ the nature of the weathered rock, granular or clayey;

○ soil type, recharge being higher on freely draining 'red' soils than on 'grey' duplex soils overlying shallow impermeable clay;

○ seasonal fluctuations in water table and corresponding yields of wells;

○ drainage pattern – fault lines usually weather more easily giving rise to depressions, often channels marked by more prolific vegetation;

○ the occurrence of water courses with alluvial beds aligned on fault lines which concentrate runoff and allow recharge to move into the fracture systems associated with the faulting;

○ the extent of fracturing of hard rock and its relation to topography;

○ the occurrence of a break in slope;

○ the strike, dip and lithological units in metamorphic rocks;

○ the occurrence of dykes, fractured or well-jointed quartz veins, pegmatite veins, chert, etc. in the area and whether they act like a groundwater barrier or conduit. Dykes are generally marked by dark soil tones and a tree line cutting across valleys and ridges alike. Where dolerite dykes lie perpendicular to the groundwater flow system the upstream contact and associated jointing may constitute a favourable site;

○ correlation between successful wells, lineament density and distance to lineament intersections on aerial photographs and to darker tones on Landsat TM images;

○ successful or failed wells or boreholes which do not fit the conceptual model of groundwater occurrence in the area; such incidences may indicate lateral variations or discontinuities in the strata;

○ the sandy or rocky nature of a stream or river bed;

○ the direction of the stream with respect to strike and dip of strata;

○ the possibility of attracting seepage from a stream, irrigation canal, river, percolation tank, contour bund, small dam, etc. towards the proposed new wells;

○ the possibility of meeting deeper permeable horizons;

○ shifting and meandering of a stream or river and evidence of rejuvenation;

○ location and discharge from previous or existing natural springs;

○ variations in quality of groundwater in the direction of its flow;

○ patches of concentrated vegetation and the likely source of its water;

○ aspect – weathering is more advanced at the base of the windward, rain-bearing side of solid rock outcrops or inselbergs.

○ large ant-hills; certain tree species including fig trees; quartz rubble on the ground surface; change in soil colour from red clay to sandy loam.

with steel or PVC to prevent collapse. Casing simplifies construction and lowers cost compared to screening, but denies direct access to water stored in the regolith. Where interconnected fractures are intercepted they allow deep boreholes to draw on the higher storativity of the regolith (Figure 2.4), and if major water-bearing fissures are intercepted some very high yields (>100 m³ d⁻¹) can be achieved. However, the pattern of fracturing in crystalline bedrock is highly variable and not easy to predict. Despite the use of a variety of geophysical techniques, many boreholes do not intercept fracture systems sufficiently well to satisfy either the volume of water required or longevity of supply. Poor success rates reflect this difficulty (Table 1.1).

The limitations of current geophysical techniques to site deep boreholes are summarized by Greenbaum (1987):

> In basement areas, fracturing provides the main source of permeability and is also important in controlling weathering and thus the development of the regolith...Remote sensing in these areas has demonstrated the existence of complex patterns of lineaments, representing faults, joints and dykes... In many areas, geological maps and reports are available which, in combination with the remote sensing data, may be used to construct a general picture of the regional structural development... These lineaments can provide a useful guide to groundwater occurrence. However, due to the limited storage potential of most fractures, sustained production at high flow rates requires in addition a zone of storage in the vicinity. Favourable conditions occur, for example, where a fracture underlies thick, saturated regolith. Such buried fractures appear often as inconspicuous lineaments on aerial photographs. Nevertheless, these are more likely to be productive than other more prominent lineaments occurring in areas of predominantly outcrop... Some comment should also be made regarding the width and nature of fracture zones within the near-surface saturated zone. Generally speaking, even large faults tend to have rather restricted zones of brecciation, typically less than a metre or two across, so it may be that accurate siting is critical... Geophysical surveys may, under favourable conditions, help to identify the weathered, saturated fracture zone and map its extent. However, if fractures offer a target as narrow as that suggested here, many techniques probably do not possess the necessary spatial resolution to accurately locate them.

Geophysical siting of wells

A comprehensive review of reconnaissance geophysical surveys for well sites was undertaken from 1984 to 89 by the British Geological Survey in Zimbabwe, Malawi and Sri Lanka. In Zimbabwe, an extensive programme of electrical resistivity profiling, combined with exploratory drilling, was undertaken. The most important conclusion drawn was that resistivity mapping could be used to exclude large areas as being unprospective, on the basis that high resistivity values represented shallow bedrock. However, it proved very difficult on a local scale to pick out suitable well sites. This was

Box 26: Geophysical siting techniques

A variety of geophysical methods (most commonly electrical resistivity traversing and depth sounding, moving coil electro-magnetic profiling and seismic refraction) have been used in a wide range of hydrogeological environments to assist in the selection of optimum sites for borehole drilling. Indeed, in areas underlain by basement rocks, considerable reliance has been placed on such methods. Varying degrees of improvement in 'success rate' have been claimed, but the published cases of comprehensive and thorough post-drilling comparison of geophysics and drilling success are few, and comparisons of different methods under the same conditions even rarer. It is, for example, not often easy to evaluate the additional benefits of geophysics, as it is rarely possible to compare on a site-by-site basis the 'with' and 'without' geophysics drilling result. In most rural water supply projects either one or other method is employed, and the best that can be expected is overall statistical comparisons of similar projects using different approaches in the same region.

Geophysical methods have been deployed in basement areas with three main objectives: a) to eliminate negative sites with hard bedrock at very shallow depth rather than waste time on unproductive drilling (experience shows this can be achieved by a variety of geophysical methods); b) to locate simply, quickly and cheaply the precise spot where the weathering is of maximum depth, relatively permeable, and where the water table is relatively shallow (this is much more difficult to achieve); and c) to locate productive fractures in the unweathered bedrock (again very difficult to achieve). It should be remembered that, to be effective, geophysical methods need strong contrasts in the physical properties of rocks, and need these to be closely related to ground water occurrence. Basement rocks are highly variable in both weathered and unweathered states, and many of these variations are gradational and small in scale. A great deal of effort over many years has been expended by the hydrogeological community in the development of instrumentation and field methods to address the three objectives outlined above. Although promising results have been reported, poor drilling success rates continue to suggest that methods that are scientifically effective and easy and cheap to operate are still not yet widely available.

due in part to the poor lateral resolution of the resistivity technique in comparison to the high degree of lateral variation within the regolith, and to problems of reliable data interpretation in unfamiliar settings.

In Malawi, no simple relation could be established between electrical resistivity and type of weathering or bedrock that occurred at a given site, and seismic profiling proved less definitive than had been expected. Resistivity soundings also gave erratic results and there appeared to be marked variations in the physical properties of the bedrock as well as irregularities in its depth. In Sri Lanka, electrical resistivity, electromagnetic, and seismic refraction data were correlated against test drilling at sites with depths to bedrock ranging from 4 to 11 m. The findings were similar to those from the African work in that results gave a good qualitative indication of larger scale variations across sites and between different localities, but the detailed structure of the weathering and precise depths to hard rock could not be predicted reliably. Depth estimates were typically no better than 20 per cent and some gross discrepancies occurred, due either to incorrect interpretation of the data or to spatial variability near the test hole.

Siting productive water points by test drilling

Based on this evidence, siting of new water points in basement areas should not rely on water divining, remote sensing or geophysics to look for 'good' fracture sites, but should focus instead on review of local groundwater conditions followed by test drilling using a portable drilling rig or power auger to directly locate 'good' storage sites where the depth of saturated weathering is greatest and permeability reasonable. Figure 5.1 shows the steps taken in evaluation of prospective areas by this approach. Water divining, remote sensing and geophysics are shown as optional, and may still help in some areas, for example, where community participation is served by divining, or where thin weathering is known and a rapid geophysical survey could usefully indicate good reason to reject certain areas and reduce the cost of test drilling.

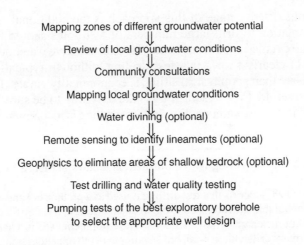

Figure 5.1 Technical steps in hydrogeological evaluation for new source construction

93

Mapping zones of different groundwater potential

The purpose of an initial broad-scale study is to define areas of differing hydrogeological potential, based primarily on rainfall, parent rock mineralogy, tectonic history and surface morphology (Chapter 2). These factors combine to determine the weathering profile and relative position of the water table, which in turn determine the most appropriate exploration and development strategy. At a regional scale, for example, weathering is better developed on the African Surface than on the Post-African Surface. Similarly, biotite is susceptible to weathering, so micaceous gneisses weather deeply, as do quartzo-feldspathic rocks with high fracture density. A map of parent rock type and mean annual rainfall thus becomes a valid projection upon which the various groundwater conditions of a region can be anticipated. Table 5.1 provides an example of zones of different groundwater potential in the driest parts of Zimbabwe, and the development strategies likely to be most cost effective in each.

Test drilling

Test drilling to site water points is not new. Clark (1996) defines a test well as a borehole drilled to test an aquifer by means of pumping tests, and illustrates the concept with an example in which five test holes are drilled in sandstone and alluvium to site a water point to supply an abattoir and associated holding pens. The use of test drilling in rural water supply, however, is not common. In Southern Africa, our test drilling has been used to locate the greatest depth of saturated weathering in the spatially highly variable conditions typical of basement. Similar use elsewhere appears limited to Sierra Leone (Box 27).

The obvious disadvantage of this approach is the additional capital cost, and the temptation to drill large numbers of test holes should be resisted. However, the detailed cost/benefit analysis of productive water point options (Appendix 1) clearly shows that test drilling, although typically 12 times more expensive than geophysical siting, is economically viable. The reasons for this are fivefold. First, test drilling is the only way to be sure of the local ground conditions and water quality, and test drilling is recommended anyway

---

Box 27: Test drilling with hand-held machines in Sierra Leone

Akiwumi (1987) reports that test drilling is used as a prelude to hand-digging wells at favourable sites confined to the weathered zone and valley fill deposits in basement rock areas of Sierra Leone. The equipment used is a hand-held mini air-rotary drill with fish-tail bit. Maximum depth of penetration is 20 m, and test drilling is used where the dry season water table is encountered within this depth.

Table 5.1   Zones of different groundwater potential and anticipated development strategies in the driest parts of Zimbabwe

Zone	Description	Groundwater potential	District and Wards	Anticipated development method
1a	Older gneiss complexes with mean annual rainfall < 600 mm	Low to moderate (2 schemes per ward)	Gwanda: Halisupi, Manama, Kafusi, Gunkwe, Ntalali, Sengezani, Buvuma, Simbumbumbu, Silonga, Mtshazo, Mzimuni, Matshetshe, Shake Mberengwa: Mutyabadza, Zvomukonde, Bhinya Road	50% SRB and CDB, 25% LDW and CW, 25% Small dams
1b	Older gneiss complexes with mean annual rainfall > 600 mm	Moderate to high (4 schemes per ward)	Shurugwi: Wards 8, 9, ... Zvishavane: Guruguru, Chionekano	50% SRB and CDB, 50% LDW and CW
2	Younger undifferentiated gneisses with mean annual rainfall < 600 mm	Low to moderate (2 schemes per ward)	Beitbridge: Dendele, Siyoka I & II Gwanda: Halisupi, Hwali, Buvuma, Manama, Silonga Mberengwa: Mketi, Mukuwerere, ...	60% SRB and CDB, 40% LDW and CW
3	Younger intrusive granites with mean annual rainfall < 600 mm	Very low (1 scheme per ward)	Gwanda: Gwaranyemba, Lushongwe, Sizeze, Simbumbumbu, Mtshazo Mberengwa: Bhinya Road, ... Zvishavane: Shauke, Mapirimira, Runde, Vukusvo, Ngombeyebane, Ture	40% SRB and CDB, 10% LDW and CW, 50% small dams
4	Paragneiss with mean annual rainfall < 600 mm	Moderate (3 schemes per ward)	Beitbridge: Siyoka I, Siyoka II, Mtetengwe I & II & III, Diti I & II, Chipise	80% SRB and CDB, 20% LDW and CW
5	Karoo basalt with mean annual rainfall < 600 mm	High (5 schemes per ward)	Beitbridge: Maramani, Machuchuta, Masera, Siyoka II, Dendele, Mtetengwe I, Gwanda: Hwali	50% SRB and CDB, 50% LDW and CW

SRB, CDB, LDW, CW: screened regolith borehole and conventional deep borehole, large-diameter well

to confirm the provisional interpretation derived from a geophysical survey. Second, where test drilling is positive a 100 per cent success rate is guaranteed. Third, test drilling does not need expensive geophysical equipment or trained geophysicists often lacking in developing countries. Fourth, an advantage of using a small test rig, apart from lower costs compared to a large rig, is that its sensitivity to the hardness of the formations penetrated is close to the experience of well digging that may follow and hence is an aid to planning.

Fifth – and most importantly – water points sited by test drilling are proving to be reliable: there has been no subsequent 40 per cent failure rate in drought as occurs with conventionally sited water points, and the higher capital costs (where these are incurred) are being outweighed by the economic benefits of reliable yield and production which accrue over time (Chapter 3).

Test drilling should be undertaken in the dry season when water levels are at their lowest. The siting of test holes in a situation where there is a minimum of data is difficult, but should be governed by the principle that every test hole should be drilled to provide an answer to a question – for example: Is there an aquifer? How thick is the weathering? How deep is the water table? In basement areas we have found that the initial review of aerial photographs and local groundwater conditions, combined with local drilling experience, play a very important part in deciding where and how many test holes to drill.

Using the checklist of natural indicators, a simple map can be made of existing water points, drainage patterns and other important features. Community members can draw up very meaningful maps of their own, showing all the relevant local information. Figure 5.2 shows the Muzondidya area; other information collected in consultation with the community is shown previously in Box 20. Figure 5.3 shows the location of 11 test holes subsequently drilled in the regolith to locate the optimum site for a new productive water point in this area. The drilling logs are shown in Table 5.2. The eleventh test hole was chosen for further investigation.

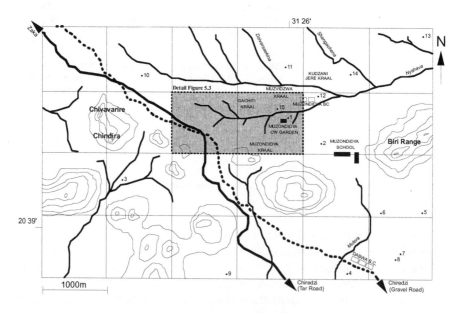

Figure 5.2 Community mapping of water points, drainage patterns and other important features in the Muzondidya area

96

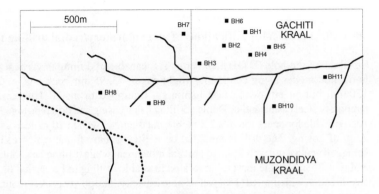

Figure 5.3 Location of 11 test holes drilled in the Muzondidya area

The test holes of 100 mm diameter can be drilled using a small down-the-hole-hammer air rig. Box 28 lists technical specifications for a combined vertical/horizontal drilling rig capable of test drilling to 40 m and radial drilling to 30 m in crystalline basement rocks, most commonly gneisses and granites. For test drilling only, or for softer formations, the specifications may be reduced and a longer mast and drill rods used to reduce drilling time.

Examples of test drilling undertaken at six sites are shown in Table 5.3, with a list of quantities shown in Table 5.4. The average cost per site in 1998 is of the order of US$1575 which compares to a typical cost of a geophysical survey of US$131. These figures are used in the cost/benefit analysis in Appendix 1.

Table 5.2 Drilling logs of test holes in the Muzondidya area

Test Hole	Drilling Log
BH1	clay to 1m, fresh rock to 5m, dry
BH2	clay to 1m, fresh rock to 3m, clay to 8m, weathered to 15m, water struck at 8m, rest water level 2.7m, poor show of water
BH3	clay to 1m, weathered to 8m, very weathered to 15m, water struck at 8m, rest water level 2.5m, poor show of water
BH4	clay to 1m, weathered to 8m, fresh rock to 12m, water struck at 8m, rest-water level 2.5m
BH5	clay to 1m, weathered to 3m, fresh to 8m, dry
BH6	clay to 1m, sludge to 3m, fresh to 6m, dry
BH7	clay to 2m, weathered to 5m, hard to 8m, dry
BH8	clay to 1m, weathered to 8m, hard/soft bands to 12m, reasonable show of water, rest-water level 1.2m
BH9	clay to 1m, weathered to 9m, hard/soft bands to 15m, reasonable show of water, rest-water level 1.5m
BH10	clay to 1m, weathered to 8m, dry
BH11	clay to 1m, weathered to 7m, hard/soft bands to 15m, reasonable show of water, rest water level 0.9m. on going deeper, hard to 31m with soft bands at 17m, 20m and 24m, good show of water in the bands, change in colour of chippings from 31m to 40m with soft bands at 35m and 39m, very good show of water in bands especially at 35m.

Box 28: Technical specifications of an exploratory/radial drilling rig

Small down-the-hole (DTH) hammer air rig, capable of drilling in vertical and horizontal modes using continuous flight auger, 73–95 mm rock bits, or 70–150 mm button bits with DTH hammers. Equipped to install 100 mm ID temporary steel casing and/or Duplex drilling using casing and drill pipe together. Also suitable for operation in a 2.2 m internal diameter well to 20 m deep using 25 m of hosing. The drill is mounted on a circular base frame which can be spragged against the well walls to prevent movement while drilling horizontally and which enables the mast to be rotated to enable drilling to be carried out in any direction with limited (15%) angle hole capacity. The power pack is mounted to the carrier vehicle. The control console is a free-standing unit containing all the controls for the hydraulic system, enabling drilling to be carried out, and monitored, from one position. It can be lowered down the well when drilling horizontally. Highly portable, the rig can be mounted on a twin axle trailer and towed behind a suitable 4×4 vehicle, or mounted on a suitable 4×4 carrier truck, or lifted on/off a larger 7–10 tonne 4×4 truck which carries all other components complete with hydraulic loading crane, stabilizer feet, 1 tonne hydraulic winch and 25 m of wire rope to lower the rig into the well for radial drilling.

○ Rotation: 0–100 rpm infinitely variable
○ Torque: 2200–2700 Nm
○ Feed/retract: 0–2.5 tonne
○ Speed: 0–20 sec m^{-1}
○ Drill stroke: 1 m
○ Mast: overall length up to 2.15 m
○ Drill head: hydraulic motor drive with spring loaded sliding sub-adaptor for easy making/breaking of drill pipe joints in any plane. Hinged to swing clear of centre line for inserting casings. Complete with water/air circulation swivel.
○ Power pack: 13–20 kW (16–27 hp) air cooled diesel engine with electric start, left on carrier truck
○ 40 m of drill rods (NWY) 0.75–1.0 m length and 50–90 mm diameter. Taper thread connections
○ Drill table: 100–150 mm diameter casing capacity, bushed to suit drill rods
○ Breakout: hydraulic ram operated to assist in breaking drill rods
○ Drill table, rotary drill head and breakout removable
○ Sub-adaptors, drive heads and handling tools
○ Compressor: typically 12 bar, 350–400 cfm
○ Foam injection facility to reduce dust hazard during horizontal drilling

Proprietary rigs requiring modifications in some cases include the TD103 by TechnoDrill, Aquadrill 60 by Atlas Copco, HPV115 Wagon Drill by Compair-Holman, Micro HR by Smith Capital Equipment and Hydroquest 2 by Demco. Cost in 1998 ranged from US$65–120 000 excluding compressor and carrier vehicle. This may be compared to costs upwards of US$200 000 for large borehole drilling rigs typically used in current practice.

Table 5.3 Test drilling at six prospective sites

Site	No. test holes	Avg. depth (m)	Days spent
Muzondidya	11	13.1	14
Gokota	4	16.3	4
Dekeza	3	14.6	4
Nemauka	4	9.5	4
Mawadze	3	17.6	4
Matedze	3	9.0	4
Average	**5**	**13.3**	**6**

Table 5.4 List of quantities for test drilling

Test drilling per site	US$ (1998 prices)
300 litres diesel	78
1 drill bit	300
6 days Driller	141
6 days Site Foreman	40
6 days Rig Hire (O&M)	1016
TOTAL	1575

Pumping tests of the best exploratory hole to select the appropriate well design

Figure 5.4 is a flow chart to help site and select new productive water points of appropriate design based on pumping tests of the best exploratory hole. Tests 1b, 1c, 3 and 4, the equipment and procedure to undertake them, and the analysis of the test data to project the safe yield that will be sustained during drought, are detailed previously in Chapter 4 and illustrated with examples in Appendix 2.

Designing productive water points for health

Groundwater in dryland areas is particularly valuable because it is often the only permanent source of water suitable for human consumption. As it moves slowly through the ground, it is filtered naturally and it can be safe to drink direct from the water point, unlike surface water which is open to numerous forms of contamination and which requires some form of water treatment. To ensure and maintain this groundwater quality, however, requires careful planning. The final location of the water point, the choice of well design

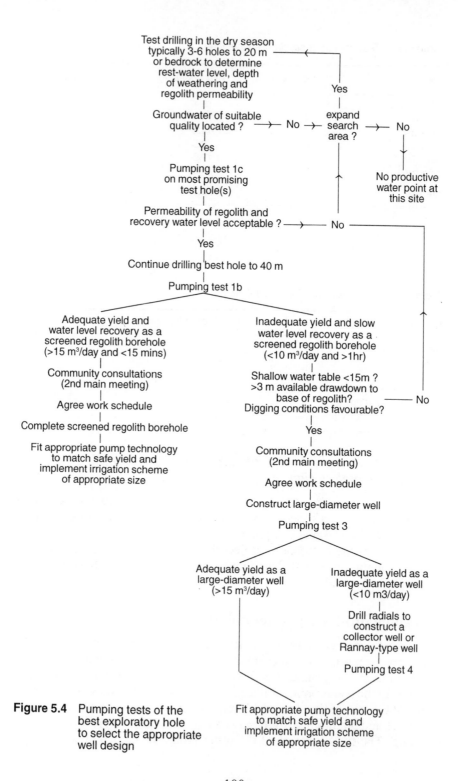

Test drilling in the dry season
typically 3-6 holes to 20 m
or bedrock to determine
rest-water level, depth
of weathering and
regolith permeability

Yes

Groundwater of suitable
quality located ? → No → expand
search → No
area ?

Yes

Pumping test 1c
on most promising
test hole(s)

No productive
water point at
this site

Permeability of regolith and
recovery water level acceptable ? → No

Yes

Continue drilling best hole to 40 m

Pumping test 1b

Adequate yield and
water level recovery as a
screened regolith borehole
(>15 m³/day and <15 mins)

Inadequate yield and slow
water level recovery as a
screened regolith borehole
(<10 m³/day and >1hr)

Community consultations
(2nd main meeting)

Shallow water table <15m ?
>3 m available drawdown to
base of regolith?
Digging conditions favourable?

Agree work schedule

No

Complete screened regolith borehole

Yes

Fit appropriate pump technology
to match safe yield and
implement irrigation scheme
of appropriate size

Community consultations
(2nd main meeting)

Agree work schedule

Construct large-diameter well

Pumping test 3

Adequate yield as a
large-diameter well
(>15 m³/day)

Inadequate yield as a
large-diameter well
(<10 m3/day)

Drill radials to
construct a
collector well or
Rannay-type well

Pumping test 4

Figure 5.4 Pumping tests of the
best exploratory hole
to select the appropriate
well design

Fit appropriate pump technology
to match safe yield and
implement irrigation scheme
of appropriate size

100

(particularly its depth), and the standard to which the water point is built, can directly impinge on the quality of water provided and the health of the users for many generations to come. These factors are considered below. Box 29 provides a checklist of design criteria to limit pollution of the groundwater resource and to avoid potential health risks.

Groundwater quality

The electrical conductivity (EC) of water from the test hole should be sampled on site at intervals during the pumping tests using a portable EC meter to detect any changes with time and drilling depth. At least two samples in 1 litre sterile bottles should also be taken for analysis for all major ions and bacteriological quality. These samples, taken towards the end of the test, will show the quality of the water and its suitability for future uses. Ideally, the samples should be kept at about 4°C by packing in ice and should be delivered to the laboratory within 24 hours. Suspended solids in the water can seriously affect analytical results, so all samples should be filtered through a filter funnel with a fine (GF/C) filter paper before being sent for analysis.

Table 5.5 lists the water quality parameters that will help you decide on potability and suitability for irrigation. Modern analyses give the results in mg/litre, which is the same as the old 'parts per million' (ppm). Be on the alert for potential problems such as salinity (high EC), high fluorides, high nitrates from agricultural fertilizers or the local geology, leachate from latrines, and runoff from livestock holding pens or other waste disposal sites. Highly mineralized groundwaters are often an indication that the rate of groundwater flow is very slow, from which it follows that recharge is low or there are no natural discharge points. Examination of the geological structures will help you decide if groundwater is trapped in this way posing a threat to long-term sustainable yield.

Bacteriological contamination

It is widely believed that hand-dug wells are more vulnerable to pollution than drilled boreholes because: the dug wells tend to be shallow and more open to infiltration of polluted surface water; their lining and surface works are commonly badly finished so that spilled water or animal wastes at watering wells can flow back into the wells; they are sometimes left uncovered allowing rubbish to fall in. Although these causes of contamination of shallow wells are sometimes observed in the field, if shallow wells are protected, our monitoring of bacteriological quality shows no distinction between shallow and deep water points. Table 5.6 shows analyses of samples taken from neighbouring community wells and boreholes after a period of heavy rainfall. In the two-layer system of regolith and fractured bedrock characteristic of basement aquifers, it appears that deep boreholes are just as prone to bacteriological contamination as shallow protected wells because both designs depend equally for their yield on shallow water stored in the regolith. More

101

Box 29: Design the water point for health

○ Sample water quality on site at intervals during drilling using an EC meter; send two samples for analysis for all major ions and bacteriological quality.

○ Where fluoride levels are high, construct shallow water points such as collector wells and screened regolith boreholes and site these further up hillsides; avoid drilling deep boreholes in valley bottoms.

○ Likewise, where salt levels increase with depth, are unpalatable or pose a threat to irrigation, expand the search area to higher parts of the groundwater flow system and construct shallow water points.

○ Boreholes appear as vulnerable to bacteriological contamination as shallow wells in the two-layer system typical of basement aquifers because both designs depend equally for their yield on water stored in the shallow regolith. To avoid contamination, both wells and boreholes should be located on raised ground away from surface runoff during heavy rain, and up the groundwater gradient (usually uphill) as far as possible from pit latrines and other waste disposal sites.

○ The upper lining must be impermeable and should cover at least the upper two metres of the water point. A concrete sanitary seal or grout is recommended. It is important to use only the minimum quantity of water, otherwise the grout will shrink on setting and not make a seal.

○ If plastic pipe is used to case and screen water points this should be proprietary purpose-made material which is safe to use for drinking water supplies. Low-cost sewer pipes should not be used because they are manufactured to lower standards and can release small amounts of toxic substances into the water supply. The best purpose-made plastic casings are made from thermoplastics.

○ The cover must be sealed to prevent rubbish and even small animals such as frogs, mice or lizards from falling into the water point, and the headworks must shed spilled water away from the water point via a concrete apron that slopes away to a concrete lined channel.

○ The headworks must encourage water to be used away from the water point. Provide outlet pipes to wash basins and to animal drinking troughs several metres from the water point.

○ The headworks must prevent effluents from entering the water point. A fence or wall must surround it to prevent livestock fouling the immediate area or damaging the structure, and the apron must be securely keyed to the lining.

○ The method used to lift water should eliminate surface spillage and human contact with the water.

○ The water point should be disinfected on completion and then annually, or as often as required during the wet season, when microbial contamination can rise.

○ During construction, participatory health education courses should be provided for the local people, including training in chlorination of the water point. Groups should be encouraged to form health clubs to oversee cleanliness and protection of the water point.

Table 5.5 Parameters to determine suitability of water for human consumption and irrigation

Parameter	Unit	Human consumption		Irrigation**	
		WHO guideline	Maximum*	All soils	Neutral to acid soils
Conductivity (EC)	µS/cm at 20°C	700	3000	700-4000	700-4000
pH		6.5-8.5	6.0-9.0	na	na
faecal coliforms	no./100 ml	zero/100 ml	zero/100 ml	100	100
Aluminium	mg/l Al	0.2		5.0	20.0
Arsenic	mg/l As	0.01		0.1	2.0
Beryllium	mg/l Be	no value		0.1	0.5
Boron	mg/l B	0.3		0.6	2
Cadmium	mg/l Cd	0.003	0.02	0.01	no value
Calcium	mg/l Ca	100		400	400
Chloride	mg/l Cl	250		100-700	100-700
Chromium	mg/l Cr	0.05		0.1	no value
Cobalt	mg/l Co			0.05	5.0
Copper	mg/l Cu	1.0	2.0	1.0	5.0
Fluoride	mg/l F	1.5	1.5	1.0	15
Iron	mg/l Fe	0.3	1.0	5.0	20
Lead	mg/l Pb	0.01	0.1	0.2	2.0
Lithium	mg/l Li			2.5	no value
Magnesium	mg/l Mg	60	100		100
Manganese	mg/l Mn	0.1	1.0	0.2	10.0
Mercury	mg/l Hg	0.001		no value	no value
Molybdenum	mg/l Mo	0.07		0.1	0.05
Nickel	mg/l Ni	0.02		0.2	2.0
Nitrate	mg/l NO_3^-	50	50	50	50
Potassium	mg/l K	10	12	12	12
Selenium	mg/l Se	0.01		0.2	no value
Sodium	mg/l Na	200		varies with crop	varies with crop
Sulphate	mg/l SO_4	250			900
Uranium	mg/l U	no value		0.01	1.0
Vanadium	mg/l V			0.1	1.0
Zinc	mg/l Zn	3.0	3.0	1.0-5.0	no value
Turbidity	NTU	5	5	no value	no value
Sodium adsorp. ratio	meq/l	—	10-15	10-15	

* Standards Association of Zimbabwe Standards (SAZS)
** *source*: Brassington (1998)

important than depth is the careful location of boreholes and wells relative to the groundwater gradient, away from potential sources of pollution such as pit latrines and other waste disposal sites.

The rate of flow of water through the ground is higher following heavy rains. With natural filtration of groundwater reduced, contamination of wells and boreholes increases as bacteria on the surface and in the groundwater are carried relatively quickly over greater distances. Water tables rise closer to the surface and to potential sources of contamination and, in the worst case, may even intercept poorly sited pit-latrines. Any defects in headworks and sanitary seals also become most obvious when water is moving over the ground surface during times of heavy rains.

The water point should be disinfected after construction, and then annually or as often as required during the wet season. To do this it should be pumped to evacuate the bulk of the contaminated water and then allowed to fill while adding sufficient chlorine in the form of calcium or sodium hypochlorite, to ensure that the water has a significant surplus of chlorine (2–20 mg l^{-1}) after dilution by the water coming back into the well. The use of granular dry chlorine, for example, comprising 70% active calcium hypochlorite and 30% inert ingredients, requires 15 g to be added per m^3 of water. For a borehole 150 mm in diameter, 30–40 m depth of water, containing about 0.2 m^3 of water, add 2–3 g. For a 2.2 m large-diameter well, 10 m depth of water, containing about 12 m^3 of water, add about 200 g. In situations where the chlorine level cannot be measured, after stirring with buckets or recycling the water with the pump, the water should smell strongly of chlorine. The water point should be left overnight if possible and then pumped to waste until the taste of chlorine has disappeared. The water should be at a pH < 8.0 and turbidity <5.0 for effective disinfection with chlorine. It is very important to use only chlorine-based disinfectants. Others such as 'pine' disinfectants are phenol based which will cause long-term taste problems and must not be used under any circumstances. Handle chlorine-based disinfectants with care. They are caustic and can burn the skin. Wear protective gloves and an overall, protect eyes with goggles or glasses, and avoid inhaling the chlorine fumes.

Fluorosis

Fluoride is a natural chemical generally found at depth in groundwater. It is associated with weathering granite rocks, and with water low in calcium. When consumed in excess, it causes painful, stiff and misshapen backs, hips and legs, and chronic gastroenteritis. In India, for example, 60 million people are now estimated to suffer from fluoride poisoning, or *fluorosis*, in areas where deep boreholes have recently been provided to supply 'safe' drinking water. It is essential that staff involved in rural water supply are aware of the potential dangers, arrange for analysis of water samples at all prospective sites during the exploratory phase of water point construction, and take evasive action where a potential health risk (from fluoride or another ion) is identified. Fluoride levels are highest in valley bottoms, and fluoride-bearing strata generally occur at depth in the weathering profile. Where fluoride poses a health risk, water points must be sited further up hillsides, and should be shallow in design. There is some progress being made in developing low-

Table 5.6 Water quality analyses of shallow and deep water points on basement aquifers in Zimbabwe (February, 1999)

Distance well to b/hole	Water point design	Geology	Water level m	Depth of well m	Faecal E.coli /100ml	Elect. Cond. µS/cm	pH	Copper mg/l Cu	Fluoride mg/l F	Iron mg/l Fe	Lead mg/l Pb	Magnesium mg/l Mg	Manganese mg/l Mn	Nitrates mg/l NO_3-	Zinc mg/l Zn
Muzondidya (0.7 km)	collector well	gneiss	2.00	15.8	60	294	7.00	<0.1	0.20	0.95	<0.1	9.19	0.1	8.9	3.96
	school borehole	gneiss	4.50	48.0	70	316	6.96	0.08	0.09	0.1	<0.1	23.69	0.2	6.6	<0.01
Dekeza (0.4 km)	collector well	gneiss	2.70	15.0	60	305	6.88	<0.1	0.13	0.20	<0.1	6.29	<0.1	8.9	2.34
	school borehole	gneiss	7.80	52.0	450	372	6.83	0.11	0.15	<0.1	<0.1	24.25	<0.1	0.9	0.6
Nemauka (0.25 km)	collector well	granite	2.40	15.0	212	145	6.57	<0.1	0.09	0.33	<0.1	17.03	<0.1	13.3	3.72
	muvavi borehole	granite	3.60	36.0	0	155	6.72	0.12	—	0.46	<0.1	8.66	0.1	8.0	1.8
Mawadze (1.2 km)	collector well	gneiss	4.50	13.0	84	638	6.90	<0.1	0.57	0.45	<0.1	22.78	<0.1	8.9	2.42
	vhudzi borehole	gneiss	17.5	39.0	960	832	6.86	0.11	0.49	2.1	<0.1	24.34	0.1	1.8	4.6
Matedze (2.0 km)	collector well	gneiss	1.00	9.5	32	331	6.94	<0.1	0.18	0.1	<0.1	59.47	<0.1	6.7	0.60
	chivamba borehole	gneiss	1.20	54.0	390	451	7.25	0.06	0.29	1.0	<0.1	22.08	0.1	0.9	1.4
Machoka (0.5 km)	collector well	basalt	1.50	9.5	18	1260	7.32	<0.1	1.01	0.22	<0.1	10.77	<0.1	132.9	1.77
	school borehole	basalt	1.25	12.0	40	1698	7.18	0.06	—	<0.1	<0.1	—	0.1	26.6	0.2
Masekesa (0.4 km)	collector well	basalt	2.60	18.0	180	581	7.75	0.09	0.4	0.4	<0.1	25.31	<0.1	66.5	0.6
	masekesa borehole	basalt	5.30	28.0	320	1425	7.37	0.07	0.74	0.5	<0.1	—	0.1	19.9	1.8
WHO guidelines					zero	700	6.5- 8.5	1.0	1.5	0.3	0.01	60	0.1	50	3.0
SAZS allowable limit					zero	3000	6.0-9.0	2.0	1.5	1.0	0.1	100	1.0	50	3.0

cost defluoridation kits for homes. Methods available include chemical precipitation through coagulation (alum and calcium hypochlorite), ion exchange, filtration through a column of crushed bones (which contain carbonates and phosphates), the use of activated alumina, and reverse osmosis. Selecting the most appropriate method requires careful consideration of costs and the reliability of the process under rural conditions in developing countries.

Arsenic

It has become apparent in recent years that groundwaters can contain natural high concentrations of arsenic, sometimes exceeding national drinking water standards. The main problem is that arsenic has not been routinely tested for in the past so existing supplies may be affected without the knowledge of the water provider or consumer. It is difficult to predict exactly which areas might be affected, although incidence in hard rock aquifers is relatively low. The largest problem so far identified is in the unconsolidated sedimentary and alluvial deltaic sediments of West Bengal and Bangladesh, where it is estimated that the number of contaminated wells probably exceeds one million. Other documented cases are in southern Taiwan, Chile, Argentina, Mexico and China. Skin diseases are the most typical symptoms of chronic exposure to arsenic in drinking water, including pigmentation disorders and hyperkeratosis on the palms and feet.

Arsenic is present in many rock-forming minerals but is most concentrated in sulphide minerals. Hence, the distribution tends to reflect variations in the type of sediments and the past and present groundwater flow. Oxidation of sulphide minerals releases arsenic into solution, so high-arsenic waters are generally found in mineralized areas and pyrite-rich sedimentary aquifers, especially where overpumping of groundwater has taken place, since falling levels introduce oxygen to the water-bearing strata and induce oxidation. As most wells have been drilled only in the last 10–30 years, the manifestation of arsenic-related diseases is relatively recent. There has been some media backlash against the use of groundwater, and in West Bengal and Bangladesh there have even been calls to abandon its use completely. However, it must not be forgotten that most wells are not contaminated and there is no reason why in uncontaminated areas the benefits that exploiting groundwater has brought should not continue. A widespread return to the use of surface water without treatment could also have its own human cost in the resultant rise in the prevalence of serious diarrhoeal disease. In developing countries, treatment methods are rarely an option. Vulnerable aquifers should be managed prudently to reduce the potential harmful effects of sulphide-mineral oxidation on water quality. Provision of good-quality water without requirement for treatment is a preferred and more cost-effective option. By investigating the local geological and hydrogeochemical environment, water quality problems can be understood better and optimum siting and design of the source rather than treatment of the symptoms becomes possible.

Salinity

If the water is for use in irrigation, the quality is important. Some salts, particularly when present in high concentrations, may harm plant growth by reducing the amount of water they can take up. Others can cause harm by their toxic effect. It is important that the water intended for irrigation use is analysed and, if in any doubt, advice on its suitability is obtained before investing in the irrigation system. Table 5.5 gives some guidelines on irrigation water quality. It should be remembered, however, that the suitability of water for irrigation greatly depends on the relative need and economic benefit that can be derived from irrigation with the water, compared to the best forgone alternative. For many people in dryland areas, groundwater is the only source of water for long periods of the year, and moderately saline groundwater is still a very valuable resource. There are many examples of waters of different compositions with salinity up to 8000 μS cm^{-1} (6000 mg l^{-1} total dissolved solids), that are being used productively for irrigation in numerous places throughout the world.

Salinity generally increases with depth of drilling (Table 5.6). It can also be spatially highly variable. Table 5.7 shows variation recorded in test drilling within 3 km^2 in a transition zone from paragneiss to younger, undifferentiated gneisses in south-east Zimbabwe. Where salt levels are unpalatable or pose a threat to irrigation, the search area should be expanded and shallow well designs chosen. The horizontal boreholes of a collector well, for example, can be used to skim relatively fresh water from the upper layers of aquifers where salinity increases with depth.

In dryland areas of subdued relief, with flat hydraulic gradients, the slow rates of groundwater movement coupled with the high evaporative losses from deep rooting vegetation, mean that there is also a more or less marked

Table 5.7   Variation in groundwater salinity recorded within an area of 3 km^2

Date	Test well	Depth (m)	Water level (m)	EC (μS/cm)
Before rains				
20/11/92	1	20	7.5	10 290
21/11/92	2	17	6.5	4680
22/11/92	3	20	8.0	10 000
07/12/92	4	9	8.1	600
08/12/92	5	8	dry	——
After rains				
18/12/92	2	17	6.3	550
05/01/93	2	17	2.9	250
05/01/93	Well A	---	4.4	7700
05/01/93	Well B	---	5.5	1120
05/01/93	Well C	---	5.9	3920

down-gradient increase in salinity which must be taken into account when siting water points. Groundwater of potable or marginal quality may be obtainable only in the higher parts of the flow system. It is an unfortunate fact that in such areas the effective thickness of the aquifers may be much less than in the lower lying areas. A compromise has to be made between obtaining lower yields of reasonable quality water higher in the landscape, or higher yields of poor quality water at lower elevations. Again, test drilling to identify the greatest depth of saturated weathering, and selection of shallow well designs such as a screened regolith boreholes and large-diameter wells, will ensure effective exploitation of these thinner aquifers.

Construction guidelines

Figure 5.5 illustrates a screened regolith borehole, large-diameter well, and collector well sited in the weathering profile typical of basement aquifers in dryland areas. The following sections provide brief construction details, including lists of materials required and relative advantages and disadvantages, of each design compared with a conventional deep borehole cased in the regolith. For more information on well and borehole construction materials, screen and gravel pack selection and installation, and borehole restoration to counter encrustation and corrosion, reader should refer to Brassington (1998) and Clark (1996).

Screened regolith borehole

Basic design: A screened regolith borehole is sited by test drilling to locate the maximum depth of saturated weathering of reasonable permeability. It is completed in the regolith and saprock/bedrock interface, typically to a depth of about 40 m. The top few metres of unconsolidated overburden are cased, and screen and gravel pack are placed to the base of the regolith. This is usually a 100 mm internal diameter screen placed in a drilled hole of 150–200 mm diameter back-filled with gravel. The lower bedrock portion is left as an open hole or casing and screen are placed to the total depth with gravel pack to stabilize the formation as appropriate.

Advantages: Being relatively shallow, a screened regolith borehole can be drilled using light to medium-weight drilling rigs. Cheaper plastic materials can be used at these shallow depths and there is a reduced head loss and pumping drawdown compared to the vertical flow to open sections in a conventional deep borehole. Loss of capacity and yield, and maintenance costs associated with clogging, collapsed zones, sand intrusion and 'sand pumping' common with poorly designed deep boreholes are overcome by the right design of gravel pack and correctness of its emplacement.

Disadvantages: The more complex design is more difficult to construct than a conventional deep borehole, and needs site supervision to select the gravel pack, base of the screen, etc. unless the contract driller is experienced and reliable.

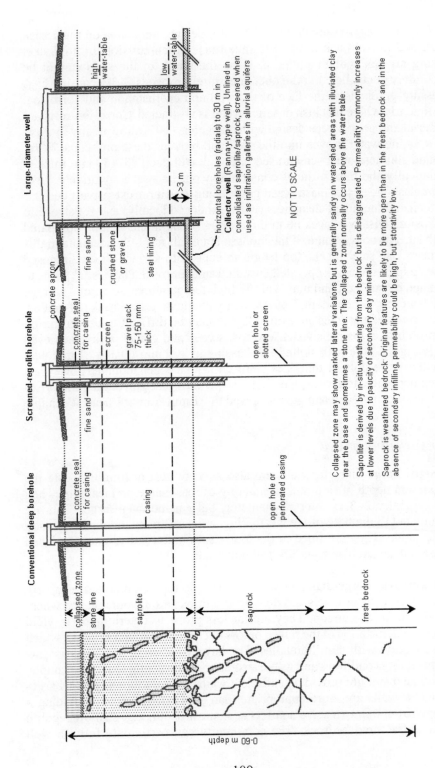

Figure 5.5 Construction guidelines for water points within the weathering profile typical of basement rocks

109

Notes on construction: Screen type, slot design, open area on the screen face, and gravel pack design are important. Torch-cut slots in mild steel casing are not recommended because the width of the slots cannot be accurately controlled and the open area rarely exceeds 2–3% causing high pressure gradients at the face of the well and encrustation and corrosion, particularly where brackish or saline water is abstracted. Corrosion-resistant material and good screen design with open area >10% allow relatively free-flow as the water moves into the well and reduces these problems. Wire-wound stainless-steel screen used with a gravel pack 75–150 mm thick is technically the best form of construction but its cost is relatively high. Fibreglass and slotted plastic casings are being used more extensively because of their lightness and cheapness (when produced locally) and resistance to corrosion. Disadvantages lie in the limited strength and rigidity compared with steel although some of the modern materials are fully adequate in this respect. Use steel for the top length to cope with surface shocks imposed when pumps are being operated, installed and removed. Plastics are prone to damage by sunlight and may warp in high temperatures, so careful storage and handling is important. Commercially available gravel packs of graded, more rounded particles are available, but may be difficult to obtain locally or prohibitively expensive. A low-cost screen and gravel pack is achieved by lining the hole with slotted PVC casing of 100 mm ID and >1 mm wall thickness, the slots <1 mm thick being cut with a bench mounted power saw at regular spaces to give an open area >10 per cent. The gravel pack comprises course siliceous sand and gravel, sieved to >1 mm from the nearest suitable river.

Large-diameter well

Design: Large-diameter wells are also sited by test drilling to locate the maximum depth of saturated weathering of reasonable permeability. They are typically 2–3 m internal diameter, but can be even larger, and are completed to the base of the weathering, at least 3 m below the dry season water table. They are lined to the total depth and a gravel pack is placed outside in an annular space 75–150 mm thick.

Advantages: Large-diameter wells are simple to construct and relatively cheap where water levels and digging conditions are favourable and where lined with brick caisson. They are far less prone to deterioration in yield through encrustation of the screen or loss of capacity through siltation, which are common with slim boreholes. The large well storage is ideal for low permeability conditions and where abstraction is discontinuous, for example, recharge overnight providing sufficient water for the following day. Large-diameter wells are more appropriate than slim boreholes for providing a supply from thin extensive aquifers at shallow depths e.g. shallow regolith over unfractured bedrock. The community can participate fully in well construction and this helps to promote local ownership of the resource. The shallow depth, reduced drawdown on pumping, and ample space to fit multiple hand or foot pumps, all mean less wear-and-tear on the pumps, less energy to lift water, and easier pump removal for maintenance by the local people.

Disadvantages: Large-diameter wells are relatively slow to construct, particularly where lined with brick caisson, which places increased demands on support staff servicing the digging team.

Notes on construction: Large-diameter wells can be dug manually or using a small compressor (5–10 bars, 70–110 cfm) to power a pneumatic pavement breaker (jackhammer) and de-watering pump (centrifugal or diaphragm-type sludge pump) which allow construction to the full depth below the water table. The final depth is controlled by weathering, digging stopping when bedrock is reached, typically at depths of 10–20 m (Table 5.8). Large-diameter wells are not yet common in Africa but are important elsewhere, for example, as 'agro-wells' in Sri Lanka and India. In Zimbabwe, they have been constructed by local communities working under a site foreman. Using shovels, picks, jackhammer and de-watering pump, a team of 4–5 men work shifts of about an hour, two men down the well at one time. The spoil is removed in a 60 litre mining kibble lifted on a 6-8 mm wire rope and manual winch and pulley securely mounted to a gantry over the well. Access to the bottom of the well is via a personnel frame lowered and raised on the winch and by steps set into the side of the well. Large-diameter wells may be lined with rocks, masonry, concrete or corrugated steel (2–3 mm thick bolted nestable round pipe). The latter offers the simplest approach but is more expensive (Table 5.9). Steel sections ranging in number from 2 to 6 and typically 0.5–1 m in depth are bolted together above ground to form a ring as the tower of casing is progressively lowered by digging out beneath. On completion of digging, the outer cavity between the lining and the open hole is back-filled with 10–15 mm diameter crushed stone to a depth of 2 m below ground level. The top 2 m is filled with concrete to make a sanitary seal. A concrete area of 4 × 5 × 0.1 m is laid around the well. A five-course brick wall is built around the casing that protrudes above ground level and this is plastered. A concrete well lid with lifting handles can be cast on site in two halves of 0.1 m thickness. The two halves should include a comprehensive steel reinforcing structure that incorporates the multiple sections of 100 or 150 mm steel casing used to mount the required number of hand or foot pumps. It is important that the stresses imposed by these pumps are carried by the steel reinforcing and not by the concrete, to avoid damage to the well lid. Alternatively, the pumps can be mounted on lengths of casing set deep in the outside concrete sanitary seal, where this is adequately reinforced.

Table 5.8   Large-diameter well construction at six sites

Site	Depth (m)	Days spent
Muzondidya	15.8	84
Gokota	15.0	85
Dekeza	15.0	80
Nemauka	15.0	116
Mawadze	13.0	83
Matedze	9.5	96
Average	**13.9**	**91**

Table 5.9   List of quantities for a 2.2 m I.D. large-diameter well to 15m depth

2.2m large-diameter well to 15 m depth	US$ (1998 prices)
Dug well construction	
2800 litres diesel	596
10m³ crushed stone (10–15 mm)	192
20 litres hydraulic oil	14
miscellaneous (picks,boots,overalls,gas, etc)	34
91 days Site Foreman	484
91 days Community Labour, 4–5 ppd	voluntary
small compressor O&M @ 2.5% capital cost	436
jackhammer O&M @ 5% capital cost	106
sludge pump O&M @ 50% capital cost	<u>270</u>
	2139
Lining and headworks	
brick caisson @ US$69 per metre *	1038
310 bricks	24
5 m³ river sand	34
33 bags cement	<u>123</u>
	1219

* 2.2m diameter by 2.0 mm thick corrugated steel lining costs US$129 per metre depth

Collector wells (or Rannay-type wells)

Basic design: A collector well is identical to a large-diameter well, but with 4–6 horizontal boreholes of 70–100 mm diameter drilled radially outwards at about 0.75 m above the base of the well to a distance of up to 30 m (Table 5.10). The design has the effect of increasing the effective diameter of the well and therefore its yield (Box 30). The dug-cum-bore well is a particular version of a collector well in which a borehole drilled into bedrock, but giving an inadequate yield, is improved by the installation of a dug well on top. In this case, sufficiently shallow water tables, diggable formations to the required depth, and the recovery response of the borehole, have all to be taken into account in deciding whether to adapt a low yielding borehole in this fashion.

Table 5.10 Radial drilling at seven sites

Site	No. radials	Length (m)	Days spent	Diesel
Muzondidya	5	15, 30, 30, 30, 30	6	1100 litres
Gokota	4	30, 30, 30, 30	6	1200 litres
Dekeza	5	8, 9, 25, 27, 28	10	1200 litres
Nemauka	4	18, 30, 30, 30	5	1200 litres
Mawadze	4	14, 16, 28, 30	5	1400 litres
Matedze	5	2, 4, 8, 23, 30	9	1400 litres
Masekesa	4	9, 19, 27, 27	5	1200 litres
Average	**4.4**	**22.8**	**6.6**	**1243 litres**

Advantages: A collector well offers the same advantage as a large diameter well listed above. It also offers special advantages in providing access to thin zones of high permeability which may be horizontal, such as stone lines and basal saprolite, or steeply dipping features such as quartz veins and fractures, whether in the bedrock or regolith. The design is also useful for sites adjacent to rivers and thin alluvial aquifers, where the horizontal boreholes can be directed beneath the river to induce recharge from the river – thereby increasing the yield. This feature also reduces water treatment costs by filtering the induced recharge through the natural filter of fine-grained sand. The horizontal boreholes of a collector well are also useful for skimming relatively fresh water from the upper layers of aquifers in areas where groundwater salinity increases with depth, and can be used to avoid harmful chemicals such as fluoride where the fluoride-bearing rocks are present at depth (Box 29).

Disadvantages: The initial capital cost of a collector well is relatively high. This is due to the additional cost of radial drilling at about US$450 per radial (Table 5.11) and the more costly steel well lining which must be used to ensure operator safety during radial drilling (Table 5.9).

Notes on construction: The radials are usually horizontal, targeted at the high permeability zone at the base of the saprolite. They are drilled using a

Table 5.11 List of quantities for radial drilling

Radial drilling (average of 4.4 radials)	US$ (1998 prices)
1243 litres diesel	326
1 drill bit	302
7 days Driller	165
7 days Site Foreman	47
7 days Rig Hire (O&M)	1195
TOTAL	**2035**

Box 30: Improvement in yield by radial drilling

In southern Africa, radial drilling at the base of large-diameter wells has improved well yields by an average of 35 per cent. Aquifer heterogeneity means that improvements vary from site to site, but where successful it has the effect of reducing drawdown and increasing the rate of water level recovery following abstraction. In practical terms, it has converted relatively low-yielding dug wells into water points that can support production, and relatively high-yielding wells into water points that can support relatively large irrigation schemes. Radial drilling is particularly valuable in those difficult groundwater conditions common in basement areas where the available drawdown is limited by relatively shallow weathering above unfractured bedrock. The site at Matedze is a good example.

Site	Geology	Depth (m)	Available drawdown (m)	Large-diam. well yield (m³/day)	Collector well yield (m³/day)	Improvement in yield (%)
Muzondidya	gneiss	15.8	13.0	11.5	18.0	56
Gokota	gneiss	15.0	11.9	16.7	17.1	2
Dekeza	gneiss	15.0	7.0	26.1	34.1	31
Nemauka	granite	15.0	9.4	24.1	40.2	67
Mawadze	gneiss	13.0	8.4	22.3	34.7	56
Matedze	gneiss	9.5	4.5	12.4	18.3	48
Machoka	basalt	9.5	1.8	47.0	no radials	na
Masekesa	basalt	18.0	9.0	55.3	62.5	13
Romwe	gneiss	12.0	6.0	10.2	10.8	6
Average	---	13.6	7.9	25.1	29.5	35

Box 31: Low-cost radial drilling in family wells of small diameter

Commercially available drills used in the mining industry satisfy the following criteria:
O small, lightweight machines which are compact, powerful and easy to operate
O suited to confined mining conditions
O capable of drilling in a 1.2 m space with 0.5 m drill rods
O powered by a 15 kW positive start air motor with a built-in silencer as standard
O lightweight rod puller and bar clamp as integral parts of the machine to facilitate easier rod handling and simplify rigging and derigging
O five-gear feed options available to suit all drilling conditions
O require only a small compressor delivering 10 bars and 110 cfm
O cost less than US$3000

small, down-the-hole hammer air rig in horizontal mode (Box 28). The drill is mounted on a circular base frame which is spragged against the base of the well walls to prevent movement while drilling. The frame enables the mast to be rotated in any direction. Where favourable groundwater conditions are identified, several radials can be targeted in this general direction; conversely, where fresh rock and little water is encountered radials can quickly be abandoned (Table 5.10). The design of the radials follows the same criteria as a vertical borehole drilled in similar aquifer material. In the consolidated regolith of basement aquifers, lining of the radials with casing or screening will generally not be necessary.

In some parts of the world, such as India and Sri Lanka, there are existing large-diameter dug-wells in which radial drilling could be undertaken to improve their performance. Elsewhere, family wells and deep wells are too narrow to accept the drilling rig described in Box 28. There is scope to investigate converting these wells to collector wells using alternative low-cost, hand-operated radial drilling tools (Box 31).

Safe working practices

The major factors to be considered in manual excavations for large-diameter wells and collector wells are methods to maximize the penetration rate and to minimize the danger to the excavators. The excavation in hard rock aquifers is by shovel, pick, chisel and jackhammer, and the debris is removed from the hole by hoist. The diameter of these well designs is large enough for two men to work down the hole together, which is an advantage in terms of safety.

The well shaft has to be completed below the water table. This is done by keeping the well de-watered, using an air driven centrifugal or diaphragm-type sludge pump mounted in the bottom of the well driven from a compressor parked clear of the well at the ground surface.

Factors which should always be considered to ensure safety during well construction are:

o   A site foreman, trained in operation and maintenance of all equipment and first aid, should be resident on site during the period of construction, and responsible for site safety.
o   Individuals should never work alone on site.
o   Ideally, transport with a first-aid kit must be on hand at all times.
o   All children and animals should be kept away from the work area. A fence can help.
o   The surface around the well must be kept clear of all loose objects which may fall on to those working below; in a large-diameter well it is possible to hang a protective wire screen above those working at the base of the well.
o   The size of the well opening can also be reduced by a cover during construction to prevent any risk of workers slipping and falling in.

115

- All hoists, wire ropes, shackles and kibbles must be of good condition and adequate design, securely fastened to the steel gantry above the well, and checked daily for wear.
- The sides of the excavation should be lined with steel or concrete rings added from above as digging proceeds, or shored up with strong timbers or jacks to prevent collapse where brick caisson is to be built.
- When working below the water table, a back-up drainage pump and adequate escape ladders should be provided.
- Hard hats with chin straps, ear muffs, gloves, protective boots, overalls, goggles and masks are vital and must be worn against dust and noise by those operating jackhammers and radial drilling rigs.
- Old wells, especially if they are more than 15 m deep, may contain carbon dioxide, methane or hydrogen sulphide. Ventilate the well before climbing into it by lowering a large bundle of sacks, rags or other material down the well several times and quickly pulling it out again.
- Old wells can also be a home to snakes: proceed with caution!
- If using electrical tools ensure proper protection by a circuit breaker.

References and further reading

Akiwumi, F.A. (1987). Groundwater exploration and development in the crystalline basement rocks of Sierra Leone. In: *Groundwater exploration and development in crystalline basement aquifers. Proc. Workshop, Harare. June 1987.* Commonwealth Science Council, Technical Paper 273, vol. 2, 431-438.

Anon. (1994). *Research and development of collector well systems: Report of an evaluation study.* ODA Eng Div, ODA, London. 36pp.

ARMCANZ (1997). *Minimum construction requirements for water bores in Australia.* Agricultural and Resource Management Council of Australia and New Zealand. 86pp.

BGS (1989). The basement aquifer research project, 1984-1989: Final Report to the Overseas Development Administration. *Technical Report WD/89/15*, British Geological Survey, Keyworth, UK.

Brassington, R. (1998). *Finding Water: A guide to the construction and maintenance of private water supplies.* Second Edition. John Wiley and Sons, Chichester.

Cairncross, S. and Feachem, R. (1993). *Environmental health engineering in the tropics.* 2nd edition. Wiley, Chichester.

Clark, L. (1996). *The field guide to water wells and boreholes in crystalline basement aquifers.* Proc. Workshop, Harare. June 1987. Commonwealth Science Council, *Technical Paper 273, vol. 1,* 137-159.

Edmunds, W.M. and Smedley, P.L. (1999) *Groundwater, geochemistry and health : trace element deficiency and excess in drinking water.* British Geological Survey, Wallingford, UK.

Foster, H.D. (1992). *Health, disease and the environment.* Belhaven Press, London. 516pp.

Greenbaum, D. (1987). Structural concepts in basement hydrogeology. In: Groundwater exploration and development in crystalline basement aquifers. In: *Groundwater exploration and development in crystalline basement aquifers.* Proc. Workshop, Harare.

June 1987. *Commonwealth Science Council, Technical Paper 273,* vol. 1, 137-159.

Handa, B.K. (1975). Geochemistry and genesis of fluoride-containing ground waters in India. *Ground Water* 13 (3) 275-281.

Harlan, R.L., Kolm, K.E. and Gutentag, E.D. (1989). Water-well design and construction. *Developments in Geotechnical Engineering 60.* Elsevier.

Howsam, P. (ed.) (1990). Water wells: Monitoring, maintenance, rehabilitation. *Proc. International Groundwater Engineering Conference,* Cranfield, UK. E&F.N. Spon.`.

Huisman, L. (1972). *Groundwater recovery.* Macmillan, London.

Kent, B. and Gombar, O. (1987). Potential for developing groundwater from strata with low primary permeability in Zimbabwe. In: *Groundwater exploration and development in crystalline basement aquifers.* Proc. Workshop, Harare. June 1987. Commonwealth Science Council, *Technical Paper 273, vol. 1,* 181-193.

Lewis, W.J., Foster, S.S.D. and Drasar, B.S. (1980). The risk of groundwater pollution by on-site sanitation in developing countries: a literature review. *IRCWD Report No. 01/ 82.* IRCWD, Duebendorf, Switzerland.

Limaye, S.D. (1987). Groundwater for irrigation and domestic use in crystalline basement rocks in India. In: *Groundwater exploration and development in crystalline basement aquifers.* Proc. Workshop, Harare. June 1987. Commonwealth Science Council, Technical Paper 273, vol. 1, 127-133.

Lloyd, B. and Helmer, R. (1991.) *Surveillance of drinking water quality in rural areas.* Longman.

Lovell, C.J., Batchelor, C.H., Waughray, D.K., Semple, A.J., Mazhangara, E., Mtetwa, G., Murata, M., Brown, M.W., Dube, T., Thompson, D.M., Chilton, P.J., Macdonald, D.M.J., Conyers, D. and Mugweni, O. (1996). Small scale irrigation using collector wells: Pilot Project – Zimbabwe. Final Report. Oct 1992–Jan 1996. *ODA Report 95/ 14,* Institute of Hydrology, Wallingford, UK, 110pp.

Lovell, C.J., Murata, M., Batchelor, C.H. and Chilton, P.J. (1998) Use of saline groundwater for community-based irrigation in dryland areas of Southern Africa. *Proc.10th ICID Afro-Asian Regional Conference in Bali, Indonesia, on Use of Saline and Brackish Water for Irrigation.* Eds. Ragab, R. and Pearce, G., pp. 106-115.

Magowe, M. and Carr, J.R. (1999) Relationship between lineaments and ground water occurrence in Western Botswana. *Ground Water* 37 (2) 282-286.

Rhoades, J.D., Kandiah, A. and Mashali, A.M. (1992). The use of saline waters for crop production. *Irrigation and Drainage Paper 48,* FAO, Rome.

Smedley, P. and Kinniburgh, D. (1999) Arsenic in groundwaters of the Bengal Basin. *Earthworks* 9, 4-5.

Thompson, D.M., Lovell, C.J., Chilton, P.J. and Macdonald, D.M.J. (1995). *Small scale irrigation using collector wells pilot project, Zimbabwe: Sites Report.* British Geological Survey Technical Report WC/95/75, Wallingford, UK.

Waterlines (1999). On-site sanitation and groundwater quality. *Waterlines* 17 (4) 1-24.

Welchert, W.T. and Freeman, B.N. (1973). 'Horizontal' wells. *Journal of Range Management* 26 (4), 253-256.

WHO (1993). *Guidelines for drinking-water quality.* Second Edition. World Health Organization, Geneva.

Wright, E.P, Herbert, R., Murray, K.H., Ball, D., Carruthers, R.M., McFarlane, M.J. and Kitching, R. (1989). Final report of the collector well project 1983-1989. *British Geological Survey Technical Report WD/88/3,* Keyworth, UK.

6 In partnership
with the community

Introduction

THE WORLD BANK'S RECENT study of 121 rural water supply projects in developing countries asked 'to what degree does participation contribute to project effectiveness?'. It concluded: 'the results are clear: beneficiary participation contributed significantly to project effectiveness…'. A multitude of irrigation development studies show the same: the sustainability of rural water supply and small-scale irrigation in developing countries depends on community / farmer participation. Carter et al. (1996) extend the argument:

> Whether or not community participation is seen as desirable for wider social goals such as capacity building and community empowerment, the hard fact is that if the community does not own, operate, maintain and pay for its water supply, then no-one else will. In developing countries, governments and most NGOs simply lack the financial resources and legal and regulatory frameworks to sustain services to growing populations.

Nowadays 'top-down planning' is often cited as the main reason for failure of rural water supply and irrigation development. But while there is general agreement on the need for a 'bottom-up approach' and 'community participation', there is often inadequate understanding of just what this process entails and what is needed to ensure its success. Participation means different things to different people. Some think that the appropriate degree of participation depends on the type and scale of the project. They argue that, for some projects, it may be sufficient merely to inform the community or obtain their agreement for planned projects, while for others it will be necessary to consult community leaders or other key people in the community for input into the planning process and, for yet other types of projects, full participation is needed throughout, from planning and design to construction, ownership and management.

Experience shows that for water supply and agricultural development to have the best chances of success, a pragmatic approach is best. Different aspects of the development warrant varying levels of participation by a variety of role players (Box 32). Success depends on achieving the right balance, in a process that is demand-driven and people-centred but which recognizes

Box 32: The partners

The State: The State should create a national strategy and enabling environment within which successful projects can develop. It may provide the bulk of the capital investment, both for source development and for training programmes and wider dissemination of information at all levels.

The local authority: While the emphasis is on communities taking the initiative and responsibility for the development of their schemes, there are a number of ways the local authority (district council) can facilitate this process. Financial assistance is normally required for various aspects, from the investigation through implementation and beyond. The administrative process through which communities can apply for short-, medium- and long-term assistance, and through which tenders are awarded to contractors, must be very clear and accessible and should provide quick answers. The criteria for project approval should be well known and supported by the national strategy, yet allow for particular local circumstances. The local authority will manage the planning of capital investment by the programme and implementation of approved investments. This should include approval of scheme-level plans and transfer of ownership to the respective communities.

The line ministries: At the operational level, the national strategy should recommend policy on specialist support services that government departments will supply in response to demand from the local authorities and communities. These services may include: source identification, planning and design, contract supervision, facilitation and community mobilization, training of users in business management, operation and maintenance of their systems, pump repair, and technical advisory services in pest and disease control, marketing, post-harvest handling and irrigation water management. The delivery mechanisms and tariffs for such services should be developed through consultation with users.

The NGOs: NGOs can facilitate development at district and community levels, and are well placed to be employed on a contractual basis through the local authority to assist with institutional capacity building at village level, formulation of social contracts, constitutions and bye-laws, community mobilization and participatory planning, provision of specific training courses, contract supervision on behalf of the community, and in some cases, provision of contracting services for scheme construction or upgrading.

The private sector: Firms and private individuals at district and community levels are well placed to be employed on a contractual basis, through the local authority or NGO or the community, to provide design and construction services, pump repair services, specific training courses, agricultural inputs such as seeds, fertilizers and pesticides, to pilot marketing arrangements, etc.

The community: The users have an important role to play in taking the initiative and responsibility for the development of their schemes, to make effective use of the services provided, and to assume responsibility for ownership, operation, maintenance and payment for their water supply.

that external support can be critical. The one element common to successful irrigation schemes is that they are all farmer-managed. Applying this principle to rural water supply, it is essential that prospective users should have full participation in the planning process and in the choice of technology, as they will eventually be using the infrastructure on a daily basis. In groundwater development, it is equally essential that at some point consultation and negotiation with the community needs to be translated into an agreed strategy to provide the basis on which the outside agency, be it government, NGO or private sector, can function effectively and provide the necessary technical inputs.

To empower a community to own, operate, maintain and pay for its water supply clearly implies very close collaboration with the community and individual community members. This chapter outlines a step-wise approach to productive groundwater development in partnership with communities, and some of the issues identified as this strategy has evolved.

Overview of the planning process

Depending on the community structure, the policy of the supporting organizations, and the range of technology options that are feasible in a particular area, productive water points may be developed to serve:

- ○ individual households that will privately own, use and maintain the resource;
- ○ small community groups such as self-help, church or school groups that will jointly own, use and maintain the system, restricting its use to members only;
- ○ the whole village or larger community group that will jointly own and use the system as a common property resource, with open access for domestic water and restricted access (to 'members') for productive water, maintenance responsibilities generally being delegated to these specific members.

Figure 6.1 illustrates a step-wise approach found helpful in participatory groundwater development. No blueprint is available to set out the most suitable approach in all cases, but important features from more successful water development and irrigation projects can provide guidance. The sequence may vary, and responsibility for particular steps may be delegated through contracts, but the main concept is gradual empowerment of the users ultimately to own, operate, maintain and pay for their water supply, recognizing the following basic requirements:

- ○ the programme must be demand driven: work with communities that have a strong felt need and which place productive water points high on their priority list;

120

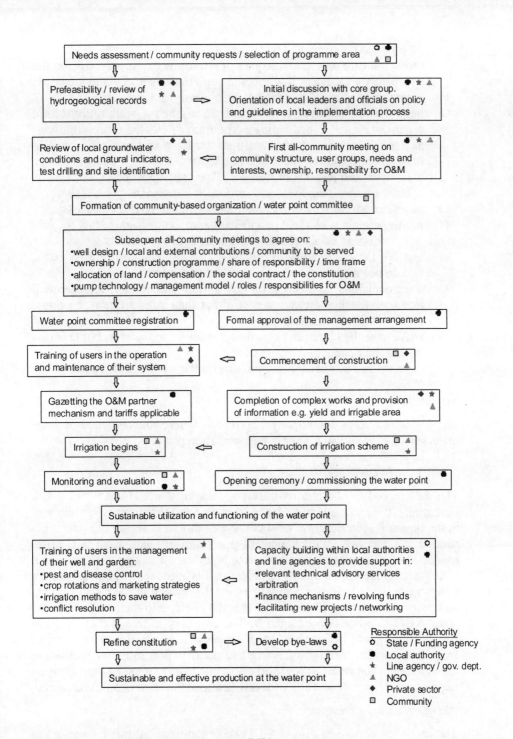

Figure 6.1 Key steps, roles and responsibilities

121

- all prospective users at a particular location must be consulted;
- provision of a safe and adequate domestic water supply will be the first priority for many low-income communities in dryland areas;
- community ownership must be established from the outset;
- help will be required to create or strengthen a community-based organization or co-ordinating committee of elected representatives, assist in defining organizational tasks, and train members in maintenance, management and financing;
- clear roles and responsibilities for all partners must be identified: the users, line agencies, local authorities, NGOs and the private sector;
- ensure room for flexibility and dialogue throughout the process;
- run the project with local people in charge, but recognize that the role of outsiders as catalysts and facilitators is often critical;
- involve the community in planning and implementing the systems, and do not present them with a pre-set plan of action;
- joint decisions are needed on appropriate technology, management and maintenance arrangements, rather than conventional top-down predetermined decisions
- make sure the community contributes funds and labour, but at levels within their capacity;
- develop and modify the programme on the basis of evaluations.

Key steps

The following notes accompany Figure 6.1, where the sequence perhaps differs from current water supply and sanitation projects.

Community ownership

The past record of community-based development initiatives is not good. Essentially, their sustainability depends on time and resources being invested at the beginning, in the form of social development to set up the structures needed to cope with management disputes and problems that will arise, especially in difficult times such as drought. Local ownership of the resource is now recognized to be a key ingredient in this approach. Communities are much more likely to look after and pay for the upkeep of their water points and irrigation schemes if they know that they belong to them and not to the state, NGO or other outside agency. The corollary of this is that the local community should be involved at all stages of the resource development, from its inception through planning and construction to maintenance and management. The state, NGO or other outside agency must be very clear on handing over this responsibility for ownership to the local people concerned.

Flexibility and dialogue

The sequence shown in Figure 6.1 is designed to avoid raising hopes too high in the community before the required groundwater resource has been

Box 33: Project initiation

Communities often need financial help to carry out projects. In this case, villagers have to meet their financiers and start to negotiate with them. It should be remembered that the resources of the community or group are the most influential factors. Local labour, know-how and investment contribute far more to the success of a village scheme than money coming from outside. Outside funds are effective only if they complement village resources.

○ Villagers believe a project is their own when they themselves have invested brain power and work in it. But a sum of money or project they get after the visit of a civil servant or nameless expert does not involve their own brain power and labour. The money or project does not belong to the villagers, and if a project is born of that one factor, the project belongs to the money-owner, not to the villagers.
○ The attitude of the villagers on such matters is often strengthened by the way the financier behaves. The financier and his experts often decide the location of the project, the way the scheme will be run, the work schedule and so on.
○ Experience shows the extent to which villagers, their customary headmen, and financiers must enter into detailed, unhurried negotiations, no matter how long it takes to reach agreement.
○ When he brings his money, the financier is often in a hurry, and thinks he can take decisions for the inhabitants. He gives them neither the time nor the means to think things out.
○ These are the usual reasons for failed projects. Failure does not necessarily come to light in the first months, when the village is full of rejoicing over the gift it has received. Failure becomes evident when relations between villagers get strained and the structures set up to deal with disputes are unable to cope to everybody's satisfaction.

Source: Chleq and Dupriez (1988)

identified. An aspect that is not adequately portrayed, however, is the need for flexibility and dialogue, particularly on the part of the financier or supporting organization (Box 33). Project initiation and project planning should be an iterative and interdisciplinary process, recognizing that interdisciplinary questions are raised throughout. The team that works in partnership with the local community must be willing and able to adopt a flexible, responsive approach and work at speeds appropriate to each particular community.

Needs assessment

Needs assessment and selection of prospective areas should be based on iterative planning (Figure 6.2). The ideal start, in terms of generating local sense of ownership, is where the community itself is able to initiate the process by making formal requests through their traditional leaders, councillors or other officials at village and ward levels. The local authority can facilitate this by encouraging requests once the development of productive water points is a recognized part of the district's development plan. It can also help by developing schemes on a ward-by-ward basis. This enables communities to see what is possible and be encouraged by initial progress, and to register their interest in the prospective project. Schemes developed on a ward-by-ward basis also allow communities to learn from each other more effectively and provide economy of scale. Factors to consider during needs assessment include:

- community enthusiasm and commitment, perhaps expressed through requests;
- the relative number and status of existing water points and irrigated gardens in the area;
- other ongoing or proposed development projects such as large dams,

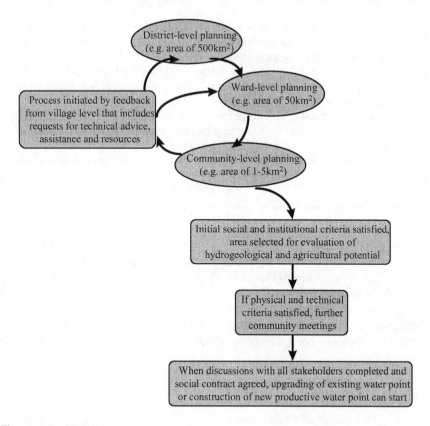

Figure 6.2 Iterative planning

irrigation schemes or resettlement programmes, their potential duplication of effort or conflicts of interest;

O   findings of the initial broad-scale study of hydrogeological potential and the viability of alternative surface water development options (Chapter 5).

First all-community meeting

The importance of community meetings in this sequence cannot be overstated. They are the key to success if properly arranged. No work should be done or meetings held in an area until local leaders and officials have been consulted. Sufficient time must be allowed for the meetings to be well organized and to be attended by all prospective users, leaders and officials. Time must also be allowed after these meetings for reflection by the community and the project staff. The first all-community meeting will focus on aspects such as:

O   the orientation of all prospective users on the policy and guidelines for implementation of productive water points *if* subsequent hydrogeological evaluation is positive. This will include initial discussions on: responsibility for ownership, construction, O&M, potential productive activities, land allocation, membership, access for different user groups, non-membership and why, participation by the poor, requirements for domestic and productive water, potential management arrangements and any tariffs for services provided, the formation of a community-based organization or water point committee;

O   participatory village mapping of community structure, number and location of households, extended families, villages, wards, roads, schools, clinics, business centres, administrative boundaries and the anticipated user-group boundaries, i.e. those likely to use the water point regularly for domestic supply and production, for domestic supply only, or for domestic supply only in times of need.

O   participatory catchment mapping of local groundwater conditions and natural indicators of water, including drainage patterns, potential sources of groundwater pollution such as pit latrines, and the location and performance of all water points, past and present;

O   planning the hydrogeological evaluation, i.e. either test pumping apparently underutilized water points in the area (Chapter 4) or test drilling to site and select the appropriate well design where new source construction is required (Chapter 5).

Subsequent all-community meetings

Where hydrogeological evaluation is successful, subsequent all-community meetings provide the venue for more detailed discussions of what can be possible in terms of well design, safe yield and productive potential if the community so wish, to confirm their commitment to ownership and O&M, to define the project community, to agree community contributions including arrangements for allocation of land for the project, to agree on the technology choice and programme of work, and to sign a social contract, usually between the local authority or NGO and the elected community representatives and officials. This is also an opportune time to initiate drafting of a community constitution by which the water point and associated productive activities will be managed.

The social contract

The social contract or document which sets out the agreement between the community and the local authority or NGO is a potent means of outlining obligations, clarifying misconceptions and enhancing community sense of ownership of the productive water point (Box 34). It can be referred to whenever there are disagreements. It should be drafted in the local language and agreed by all prior to signing in public by elected community representatives, an officer of the NGO, and a government officer as witness. Worked examples of social contracts developed for groundwater schemes and small dams are provided in Appendix 3.

Box 34: Social contract

Basis of agreement for construction, ownership, operation and maintenance of a productive water point in your area

Productive water point name: ....
Local Authority or NGO name: ....

Ownership: the community in the villages of.....

Construction:
        Project staff will provide: ....
        The Community will provide: ....

Community responsibility:
        All management decisions
        Upkeep, maintenance and repair
        Any future improvements

Suggestions to help the project be a success:
        Full participation to decide membership
        Full participation to decide land allocation
        Full participation in construction
        Collection of a small fee to buy initial inputs

Agreement: Community: ....   NGO: ....
        Government: ....   Date: ....

Community contributions

Sensitive criteria are needed to establish contribution mechanisms that take into account local socio-economic differences prevailing in the community. Local sense of ownership is best promoted by maximizing these contributions, in terms of material inputs in cash and in kind, and in the number of people involved. Where there is an ability and willingness to pay for a higher level of service, charging part or all of the economic cost to the consumer will also make resources available for a higher population coverage. However, the type and extent of contributions must be agreed in open dialogue, in a manner that does not discriminate against any particular group, and which is consistent from scheme to scheme and with the socio-economic and cultural attitudes of the people concerned.

Box 35: Community contributions

'Development is increasingly understood to be a process whereby people learn to take charge of their own lives and solve their own problems. Helping people solve their problems by giving them things and doing things for them makes them more dependent and less willing to solve their own problems. This cannot be called development: on the contrary, it is the very opposite of development.'

One way to reduce the harmful impact of subsidies is to require matching labour contributions. The idea is as follows. The group may choose their own water supply technology (drawing on external advice on available options where appropriate) and must build half the given structure with their own labour (either family or hired). They may then request the programme to construct the second half, according to the design specified by the group. The programme would hire the workers to do this under the programme; these workers would be paid only after the group certifies that the work is acceptable to them.

This approach has several advantages. First, it helps to ensure that the technology suits the group's wishes and is built according to standards that satisfy them. Second, the group never receives any payment, reducing the chances that they will participate in the programme for ulterior reasons. This approach also offers an important side benefit of helping to organize unemployed people and teaching them skills that will increase their self-sufficiency. In particular, labourers may form an association to provide water supply construction services. Payments from the programme would be made direct to the association and distributed to its members. Assistance could be provided to the association to develop their business skills and perhaps develop spin-off activities such as revolving credit programmes. More generally, this idea follows the principle of using rural development programmes as leverage to create benefits for disadvantaged groups such as the poor.

Adapted from Kerr et al., 1996

Types of contribution include cost recovery in some proportion, either as loan repayments or payment of maintenance costs, 'food for work', provision of local materials, voluntary labour and matching labour (Box 35). The people's experience of previous development projects, and the scale of the particular project, can be important deciding factors. Mechanisms such as revolving funds and local banking facilities may have to be established to ensure that payments can be made and used. Generally, paying a user group for work done is not recommended, whether by project staff or by other community members, as it promotes neither a sense of ownership nor good progress and causes problems for low-income families who struggle to raise cash. Automatic membership of the project for those who do the work is preferable.

Scheme membership and participation by the poor

Many governments and support agencies recognize that there are particularly disadvantaged groups among populations living in the marginal areas, such as female-headed households or young families without land. These groups are most in need of improvements but are often not reached because they lack capital, labour, draft-power, are less healthy, or live far from the water source. However, development that targets specific groups can create divisions within a community. Target specificity tends to confuse villagers, who generally believe in development for all. The intention therefore must be to provide a smooth transition towards better water supply and production for all. An important lesson from rural development is that this will not happen automatically. Local élites and economically more able or active households can monopolize community developments, and poorer members of the community – and women in particular – tend to be marginalized unless specific measures are taken to include them. User communities should be encouraged therefore, to develop 'enabling strategies' in the form of social contracts, constitutions and other mechanisms which ensure at least proportional representation by the poorest and most vulnerable groups. Box 36 lists a number of special steps that may be necessary to involve these groups alongside wealthier families, who are often participating more automatically.

Community planning

Although the emphasis is on communities taking the initiative and responsibility for the development of their schemes, success in productive groundwater development depends on securing the required water source. The objectives of community members and project staff are similar in this respect. Both aspire to establish a sustainable water point that can effectively be used for domestic supply and production. Both should contribute to the planning process. The community has knowledge of local conditions and can ensure that the facilities meet their own requirements. The project staff

Box 36: Steps to include women and the poor in the full project cycle

○ The introduction of water and gardening projects can be seen as a way forward to working with women and the poor. Project staff should ensure their maximum involvement throughout the development process. The collection of water and the provision of vegetables are both traditionally considered women's concerns. The empowerment of women to make decisions in water management and irrigation guarantees their taking increased responsibility in food security and household budgets.

○ Employ women as field agents and orient male project staff to the benefits that involving women and the poor have on the achievement of project objectives.

○ Agree with local leaders on the need fully to involve women and the poor.

○ Ensure all groups are informed about all meetings and that they participate.

○ Organize meetings when and where the most vulnerable groups are able to attend.

○ Organize women participants to sit together at meetings in a prominent location, and stimulate discussion by actively inviting criticisms from all groups, including women and the poor, and asking respected women to speak.

○ Select pump and irrigation technologies that are readily used by women. Consult on which management roles are best carried out by women and explain the requirements to candidates.

○ Adapt training programmes to suit the needs and capacity of the most vulnerable groups rather than the least vulnerable groups, and provide follow-up support and monitoring of progress.

○ Avoid joining fees or fixed capital or labour inputs; instead, investigate differential contribution mechanisms that take into account the local socio-economic differences prevailing in the community and which the people agree are appropriate to their own varying needs.

○ Promote equity by avoiding closed or exclusive membership: for example, by a system in which every family in the community is entitled to a set number of beds in the garden, but each year at harvest the other additional beds and any non-utilized beds become available again at a set price for newcomers or at an annual rental for members wishing to expand for the coming year.

have knowledge of the technology alternatives and the financial implications, and can advise on construction and maintenance arrangements. Building on local knowledge is the most efficient way to initiate the process, but hydrogeological evaluation to site and select the optimum well design depends on outside support if the potential of the groundwater resource is to be realized (Chapters 4 and 5). It is important, therefore, that transparency is maintained during this step to ensure that the local people remain clear on the reasons why the particular site and well design are being recommended, and the socio-economic implications of this choice in terms of cost, productive potential, land allocation and membership. An open dialogue in which the possibilities and limitations of the proposed support are explained is an essential foundation for partnership throughout the development process.

The Constitution

The constitution is a document that outlines how the water point and associated productive activities will be managed. It is a community initiative with the guidance of the NGO and local authority counterparts. The community leaders, referring to the Social Contract, remind the people that as a community they have a responsibility to honour the contract. Normally, the leaders ask the people how they can ensure that all interested people will participate in the project since they now have a contract in place. The community usually suggests that a constitution be drawn up (Box 37).

A meeting is held to discuss issues that may result in project failure. These form the basis of the constitution. NGO and local authority staff explain the importance of the constitution to sustainability, and give examples of constitutions that have been drafted by other communities. They facilitate this meeting and can provide guidelines on the drafting. The community is asked to come up with their own constitution. When ready, it is read in public and all members are free to make adjustments. The supporting agencies are allowed an opportunity to comment since they are interested parties, but may not influence changes to the constitution directly. The community then adopts the constitution by vote.

The constitution will normally provide guidance on: membership, equity and participation of the poor; financial contributions and user-fees; cost-recovery and the O&M partner mechanism; reinvestment of revenue; rights of access, particularly for non-garden members wishing to use the productive water point to start new income-generating activities; the water point committee election process; the role of the committee and period in office; treasury arrangements; dispute resolution procedures, etc. The constitution should also clarify the roles and responsibilities of the local authority and line ministries in providing institutional support.

Allocation of land for the water point and irrigation scheme

The question of who 'owns' a common property resource is compounded when that resource assumes an economic value through production. The problem is particularly acute in communally managed areas and in community-based projects that need land. Allocation of land for a productive water point is no exception. Conflict can arise where a local leader or other influential member of society may 'give' the land (and the groundwater beneath) for the project but subsequently undermines the role of the elected water point committee by refusing right of access for some user groups. It is

---
Box 37: The constitution

Name of productive water point:
Location:
Objectives of productive water point:
Objectives of Constitution:
Community boundary / Area of influence:
Membership
        Rights of access:
        Termination of membership:
        New membership:
General meetings:
        Rights of the community:
        Obligation of members:
        Voting:
        Quorum:
Water point management committee:
        Election process:
        Term of office:
        Duties:
        Powers:
User fees:
Payment for repairs and maintenance:
Banking / re-investment of revenue:
Registers and records:
New projects at the water point:
        for members of the initial project:
        for non-members of the initial project:
Settlement of disputes:

    Water point committee: ....   NGO: ....
    Government witness:........   Date: ....

---

important that formal transfer of ownership of the land for the productive water point and associated activities is made public prior to signing of the social contract. This may involve compensation being agreed between leaders of the community and the present landowner, and the process can benefit from the presence of a member of the local authority as witness and with whom the official transfer of land and agreed compensation can be lodged.

Water point construction

Construction should begin after formal allocation of land for the scheme and signing of the social contract. Part of the agreement between community and project staff should be the provision of a trained site foreman who will help the community to construct their water point and irrigation scheme where this is the chosen productive activity. The site foreman will be responsible for training community members in the use of special construction equipment, and for health and safety on site (Chapter 5). Where an existing underutilized water point is to be upgraded (Chapter 4) or where new source construction requires a screened regolith borehole or conventional deep borehole (Chapter 5), the construction phase will be relatively short and community participation relatively low, as the drilling contractor will generally be self-contained and independent. Nonetheless, it is important to maintain open dialogue at this stage so the community are equal partners in any decisions that are made. Where construction requires conversion of an existing well or borehole to a more appropriate design, or the construction of a large-diameter well or collector well, community participation in construction will be considerable, and this can be a real advantage in helping to generate good community spirit and local sense of ownership of the project.

The O&M partner mechanism

Alternative management systems for O&M are possible, in which roles and responsibilities are shared in different proportions by the community-based water point committee, the local authority, the government, and the private sector. These models include:

○ community fully responsible for undertaking O&M themselves;
○ community can call on (and pay) a government pump-minder;
○ community can call on (and pay) the local authority to commission repair;
○ community can pay a private pump-minder;
○ community can lease the pump(s) from a private company or NGO that guarantees to undertake repair (Box 17).

Choosing the model is a community initiative, as opposed to a conventional top-down predetermined decision, and the NGO and support agencies should provide guidance on the technical and financial implications of each option, including tariffs where these apply. In many developing countries there is a need to gazette a list of fixed prices, and to encourage the participation of local artisans and workshops alongside government to provide a commercial O&M service. In all models, the signing and effecting of a tripartite agreement between the community, local authority and support agency will be very important for legal and social recognition. It is also important to note that current practice, in which the local authority is responsible for O&M, is not considered appropriate for productive water points, in the light of the poor record of O&M achieved by this approach (Chapter 1). This raises again the legal issue of productive water point *ownership*. The community-based organization will need to register as a legal entity, and the transfer of assets and liability for O&M from the local authority to this organization must become possible under the existing bye-laws.

Community training in pump repair

On determining the safe yield of the water point, it is vital to discuss the future water lifting device thoroughly with the community. Too many rural water supply projects have failed because there was insufficient exchange of views. Chapter 4 provides a checklist when choosing appropriate pump technologies for productive water points.

Multiple hand- or footpumps fitted to the water point can make effective use of the potential yield, reduce queuing, and provide an important safety net for irrigation when one pump breaks down. In many cases, this is preferable to a single motorized pump which has higher running costs and difficulties of repair often beyond the means of rural villages, and which lacks the control against groundwater depletion inherent in manual pumps, an important consideration in dryland areas.

In the absence of alternatives at present, local people at productive water points in Zimbabwe have been forced to adopt either the first or the second O&M models listed. To aid them, multiple B-type handpumps (the 'standard' pump in Zimbabwe) have been used. In addition:

○ A steel gantry above each well has become an integral part of the water point design to allow lifting of the pumps for repair.
○ Training days at the time of pump installation provide 8–10 men and women elected by each community with the opportunity to dismantle and reassemble the pumps, and generally gain confidence in their ability to effect repairs.
○ The importance of routine maintenance, regular greasing and tightening of loose bolts, is emphasized.
○ Tools and an easy-to-follow repair manual are provided as standard with the pumps. The tools comprise: 2 x 450 mm pipe wrenches; 1 x 250 mm adjustable spanner; 1 x 20 m rope; 1 x 50 mm pipe clamp; 1 x 50 mm lifting plug.
○ All schemes have been registered with the relevant local government department to benefit from the third tier of the government's maintenance programme, which is supposed to provide assistance in cases of emergency.

This approach has worked reasonably well and its success and popularity is reflected in the continuing ability and willingness of these communities to repair not only their own pumps but also those of neighbours at non-productive water points. At some schemes, members are even purchasing spare parts in anticipation of future breakdowns. Confidence in self-maintenance comes with practice. However, it should be noted that daily wear and tear on even these so-called 'robust' handpumps is considerable (Box 22). The pumps do break down and the present heavy weight of some designs is a disincentive to community-based maintenance. Alternative lightweight 'extractable' handpumps should be used where these can satisfy the required pumping capacity. Purchase by the community of a SIWIL pump lifter or Pulltite and Suregrip borehole maintenance combination set can also help where pumps are particularly heavy on deep boreholes.

Completion of the productive water point

Involving local communities in all aspects of planning, ownership, management and maintenance must inevitably involve equipping these same communities with the necessary information and training. The NGO and supporting agencies should provide guidance on the safe yield of the water point and irrigable area, where appropriate. These calculations are detailed in Chapter 4. The water point, when complete, will require a fence or wall to provide a sanitary space and keep animals out. Fence design and garden layout are discussed in Chapter 7.

Official opening ceremony

Official openings of productive water points are enjoyed by all, and provide another important opportunity for community ownership of the scheme to be confirmed by all the outside agencies that might have assisted during scheme development. This event can be organized by the local authority and attended by local and foreign dignitaries, as appropriate. It helps to seal the sense of ownership felt by the community, who can show with pride what they have already achieved. It is an appropriate time for a celebration.

The project cycle should not end there, however. Participation in the development of productive water points provides the ideal springboard for other community-based projects where collective action is important, especially in health, agriculture and the environment. Villagers given a reliable source of water and income will soon begin other income-generating projects, ranging from keeping rabbits and growing fruit trees to knitting clubs and dairying. Community workers in the area will be able to advise on nutrition at the new gardens. Agricultural extension staff and NGOs will be called upon to provide advice on irrigation management and to help the community to formulate their ideas for new projects. Local authorities may be called upon from time to time to provide arbitration in disputes. Perhaps most important is the opportunity this type of water development project provides to initiate community-based management of other natural resources in the surrounding catchment. In this way, the people themselves can become responsible for recharge to their well and for the life of their scheme, and in so doing begin to address the problems of environmental degradation – often too daunting when considered at any larger scale. These aspects of productive water point management are considered in the next chapter.

References and further reading

Attanayake, M.A.M.S.L. and Jayasiriwardana, D.S.D (1996). System development for future sustainability. In: *Reaching the unreached – challenges for the 21st century. 22nd WEDC Conference*, New Delhi, India. Vol 1, 16-18.

Carter, R.C., Tyrell, S.F. and Howsam, P. (1996). Effective and transparent strategies for community water supply programmes in developing countries. In: *Water policy: allocation and management in practice*. Eds. P. Howsam and R.C. Carter. p.162-168. E&FN Spon Publs.

Chleq, J.L. and Dupriez, H. (1988). *Vanishing land and water: soil and water conservation in dry lands*. Land and Life Series. Macmillan Publishers, London.

Clayton, A. (1999). *Contracts or partnerships : working through local NGOs in Ghana and Nepal: how local NGOs have been engaged in the provision of rural water and sanitation at community level in Ghana and Nepal*. London, UK, WaterAid. ISBN 0 9513466 8 7.

DFID / WELL (1998). *Guidance manual on water supply and sanitation programmes*. Department for International Development, London. 338 p.

Dudley, E. (1993). *The critical villager: beyond community participation*. Routledge, London.

Kerr, J.M., Shanghi, N.K. and Sriramappa, G. (1996). Subsidies in watershed development projects in India: Distortions and opportunities. *Gatekeeper Series No. 61,* IIED

Lovell, C. J., Mazhangara, E. P., Murata, M. and Mugweni, O. (1999). Key steps in the development of community gardens using limited groundwater resources. In: *Water for agriculture in Zimbabwe: Policy and management options for the smallholder sector*, Eds: E. Manzungu, A. Senzanje and P. Van der Zaag. Chapter 6, p. 78-91. University of Zimbabwe Publications.

Moriarty, P.B. and Lovell, C.J. (1998c) Water resource development in Chivi District, Zimbabwe: Results of a village mapping exercise. *IH DFID Report 98/10*, Institute of Hydrology, Wallingford, UK.

Narayan, D. (1995). The contribution of people's participation: evidence from 121 rural water supply projects. *Environmentally Sustainable Development Occasional Paper Series No. 1*, World Bank, Washington DC, USA.

7 Management of productive water points

Introduction

DEVELOPMENT OF SUSTAINABLE AND effective productive water points requires the long-term commitment of the project agencies involved, with support focused primarily on capacity building within the communities and the agencies themselves. The partnership approach does not end with the implementation of infrastructure, but continues with the active participation of the community (often through an elected committee), in managing and maintaining their scheme. The users should regularly evaluate their utilization of the water point and the operational arrangements around it, to refine and improve these as their confidence and experience grows.

Providing an opportunity for community feedback to identify gaps in skills and knowledge is an important step in this process. It involves meaningful information exchange between the community and their technical advisers, with the community taking the lead and the advisers playing a facilitative role. The plans are the community's and they consult the facilitators to ensure that their plans are technically feasible. While the thrust is primarily towards self-help, the community is helped to form links with potential sources of outside assistance to facilitate positive impacts of the water point on the wider production systems.

Organizations with the capacity and understanding to listen and make meaningful input at village level will play a vital role in this regard. Government extension agencies, local authorities, contracted NGOs and private training institutions each have an important role to play in this programme of institutional capacity building and organizational development.

The effectiveness of community-based management at productive water points has been monitored in Zimbabwe for the past eight years. This has consistently highlighted the need for follow-up training and aftercare in five key areas:

○ organizational and institutional development
○ management of community gardens
○ garden design and operation
○ productive water points and the surrounding environment
○ drought management.

This chapter highlights management issues under each area, and considers the programme of support that can help communities to make more effective and sustainable use of their productive water points.

Organizational and institutional development

Different user groups

Productive water points and their associated projects should not be conceived as closed systems, and community-based management strategies to cope with the demands of different user groups must be considered. Where an initial project is attached to the water point to make use of its productive potential, three user groups, and sometimes more, can be defined within the community. These groups operate at different spatial and temporal scales and apply different pressures for their own production systems enhancement (Box 38). Group 1 members are clearly focused on the new project and its effect on their farming systems and livelihood strategies, thinking further ahead as income security increases through the project whereas Group 2 and Group 3 members are focused more on the immediate supply of water and their independent livelihood strategies, with less incentive to consider long-term management issues.

Generally, demand for new uses of the water to initiate or expand projects will come from those within Groups 1 and 2. As this demand increases with time, conflicts of interest between different groups pose increasingly difficult management problems for the water point committee. Which projects should be allowed? Should there be a bias towards one user group or another? Should new projects be for groups only or for individuals as well? How will the different projects be managed? How will water be paid for and O&M organized?

Problems emerge, for example, where initial project members complain if non-members begin taking water for other activities, or misuse or overwork the pumps, or contribute less money or labour to O&M. Where water is limiting, a flat-rate disincentive charge may even be levied to discourage Group 3 seasonal users during times of drought. People may be barred from using a water point simply because 'they come from too far away'.

New projects

Clearly, special care in setting up productive water points is justified, because they have such widespread and long-lasting impact on local livelihood and production systems. As we have seen in Chapter 6, this requires more time than has been spent in the past to allow the community members and leaders to develop their own ideas on the shape of the development and to weigh up its expected costs and benefits. Making undue haste at this stage to expedite programme implementation is counterproductive. Open dialogue and agreement is essential before scheme implementation, and every effort must

Box 38: Different user groups and their different demands

Group 1: Members of the initial project, often living in the immediate location, who use the water point for domestic purposes and for their new productive venture. It is often this group who form the committee to manage the water point and initial project. Often comprising representatives from several villages or extended families with the project as their common interest, they are able to focus and lobby their interests effectively through their committee. With the appointment of a chairperson, secretary and treasurer, this group usually contains key members of the community having experience in village-level administration.

Group 2: Non-members of the initial project but who use the water point for some or all of their domestic water and sometimes for other productive activities. Generally a more disparate group drawn from a broader section of households, this group may fluctuate and is harder to delineate because some families use other water sources as well. Unless an effort is made to include them in the management committee, Group 2 water users have much less of a coherent focus or lobbying power than Group 1. Viewpoints are expressed either by prominent individuals or through non-participation in maintenance of the water point. Pressure exists within Group 2 from people who now wish to join the productive project but who were initially absent, unable to join, or risk-shy.

Group 3: A scattered and variable population of users who use the water point on a seasonal basis, particularly in times of general water shortage. Experience shows that Group 3 water users consist of those households who do not have a reliable domestic water point closer than the productive water point and who create extra demand when the other water sources fail. Again, unless specific measures are taken to include this group in the management committee, their lobbying power will be weak and they will play little or no part in management of the water point. It appears that Group 3 users are rarely considered in any water and sanitation project.

Other Groups: Depending on the particular location of the water point, other user groups can include schools, clinics, businesses and individual entrepreneurs. Such groups can exert significant lobbying power and affect committee decisions.

be made to involve *all* who may subsequently use the water point.

Enabling each of these user groups to meet and discuss their access to the opportunities for production is critical to the equitable use and management of productive water points. Surveys have found that some new projects have set up separate management committees from the water point committee. This obviously creates a conflict of interest and poses problems for the initial committee. Examples include groups who separately planned to invest in fruit tree and seed nursery projects while the water source they would ultimately depend on was managed by the water point committee. In cases like these, members of the new projects should clarify and formalize their relationship with the existing water point committee and with other institutions in the area, including traditional and political leadership, village and ward development committees and extension services. These institutions play an important role in advising and supporting members of productive water point initiatives and can also help identify matching funds or credit facilities to start the new projects. Any expansion of production systems within the area should actively encourage the involvement of these institutions.

Where demand exists to start new projects or expand the initial project, group discussions will contend the range and size of projects that can safely be initiated. It is important that the water point committee has information to hand on the safe yield of the water point and can decide if the new initiatives are viable in this respect. A second issue likely to arise is whether initiatives conceived by non-members of the original project (Groups 2 and 3) can be developed. The decision rests on whether further development at the productive water point is geared only for those members of the initial project. If this is allowed to happen, then the production systems enhancement process will take off for some but leave others behind, resulting in inequality. Again, this comes back to the implementation procedure used to set up the scheme, the steps incorporated in the project constitution to deal with equity and new membership, and the level of village government that prevails.

The concept of non-members implementing new group projects and the legitimacy of their management committees in the eyes of the initial water point committee are seen as important issues by people living around productive water points. Where village government is weak, coherent development is difficult. In such cases, participatory development of community resource management plans is an important next step following infrastructure completion. This will help to identify the principles and community organizational structures required to co-ordinate various new projects and associated management systems for sustainable water supply and production systems enhancement.

Security of tenure

Lack of clarity on usufruct rights in respect of land and water can result in disincentive to invest in, and inequalities in access to, water points and associated productive activities. Once such disincentive and inequalities are created, they are difficult to resolve (Box 39).

Socially acceptable ways of securing land for productive water points are needed, therefore, to minimize the conflict that can surround a piece of land 'donated' for group use. Conflict arises, for example, if the previous 'owner' of the land is an influential person who then subsequently begins to override the water point committee, for example, on rights of access for Groups 2 and 3. The same problem can arise at standard domestic water points, where groups from outside the immediate village may be barred from using the resource.

Experience shows that security of tenure and rights of access are enhanced where the water point can be successfully sited on land previously used by the whole community (such as part of its grazing land) rather than land used by a particular household (such as arable land). Where groundwater conditions preclude this, other ways should be explored, such as securing land for the project through a Land Allocation Committee or getting the local authority to take steps to gazette a defined piece of land with suitable groundwater for this purpose. Security of tenure guaranteed by the local authority is generally the best solution if the community is to make long-term investments involving land development and costly infrastructure such as motorized pumps. The local authority should develop a register of the projects whose security of tenure they have guaranteed.

Box 39: Tenure in communally managed areas

Tenure in so-called communal lands presents something of a paradox. On paper, the government owns the land. On the ground, the situation is often reversed, with real power over land allocation frequently being wielded by traditional leaders. Attempts by government operatives to intervene are often ineffectual and issues of squatting, illegal cultivation, illegal land clearance and so on are often reported. A review of this sensitive question has recently been undertaken in Zimbabwe by the Land Tenure Commission. The situation remains difficult until new government policy is defined. In the meantime, community-based projects in communal lands have to find the right balance between working with formal and informal institutions, and it is vital that all institutions be consulted in the development process.

Devolution of authority

Some government actions are achieved by passing directives down from government to the people on the ground. Others need to be achieved through a partnership between government and organizations that represent grassroots stakeholders. Successful community-based natural resource management definitely falls into the latter category, with the state devolving its authority over resources to communities who become proprietors.

Government can no longer issue directives or enforce their compliance, so long as the proprietors manage the resource properly and conform with any agreement to do so. The role of government should shift from a director to the facilitator of sound management, through the provision of information and other support services such as research and extension. In effect, the state trades power for influence; a police force for an extension worker in a field where policing has been expensive and ineffective. This is because resource management is now driven by institutionalized users and group incentives, rather than attempts to enforce flawed and socially unacceptable legislation.

This devolution of authority over common resources will succeed only if the legal and economic setting is clear and conducive to success. It requires technical support ranging from how to set resource use limits for users, to mechanisms for dealing with offenders. Such issues need not be complex and are within the capacity of rural communities to implement. There is no contradiction between government or local authority being the ultimate authority for natural resources and devolving the responsibility to manage to local communities. The people on the land are in the best position to manage efficiently and should be encouraged to do so, while government or local authority monitors the process to ensure that it is sustainable.

Box 40 provides an example of devolution in Tanzania, in which water point committees, village governments and the local authority are successfully working in partnership to manage motorized water supplies. Control resides at the village level, not at the District, Division or Ward levels, which are the hierarchies in Tanzania. This is important. A problem in Zimbabwe, for example, is the lack of a clear lower-level unit below the district to which management authority can be devolved, given the friction that often prevails between political and traditional leaders at kraal, village and ward levels (Box 41).

Leadership

Experience shows that productive water points require natural leaders in the community who are willing to take the initiative and make the project a success. The step-wise approach to development outlined in Chapter 6 is designed to facilitate this, by helping the users to identify leaders among themselves. Through creation or strengthening of a community-based organization of elected representatives and drafting of a project constitution, a foundation is laid and management arrangements set in place to cope with the day-to-day problems that will arise and the issues that might otherwise cause project failure.

Organizational skills take time to develop, however, and may require some external support over a long period. Trained facilitators with the skills needed to guide communities can play a key role in building capacity for community-based management.

Technical training and support

Training should be given a high priority because it is one of the activities most likely to enhance sustainability and effectiveness of productive water points in the long term. Exposure visits to successful schemes, and group meetings between new and well established schemes, are useful ways for communities to exchange experiences, explore the various leadership styles and management options, and focus on gaps in their skills and knowledge.

Box 40: Village government and the water point committee: Tanzania

Ngomai is in Kongwa District of Dodoma region. It has a population of 3,916 and a livestock count of 1,325. Although there are some ponds and family wells, the water supply system is based on a borehole that yields 60 m^3d^{-1}, a motorized pump, a tank of 20 m^3, and a distribution network of four domestic points. The village government (VG) has 8 elected members representing people in each sub-village. Sub-committees (or ministries) supervise activities such as social services, finance and security. A village chairman is elected by the entire village. A village executive officer (VEO) is employed by the local authority to oversee day-to-day affairs. The water committee (WC) falls under the social services committee. It is elected by the community, has 16 members, 8 of whom are female, and all represent their sub-villages.

The VG requires the WC to: submit monthly and quarterly reports; audit income and expenditures; assist in security at installations (the borehole and the domestic points); enforce bye-laws on the use and protection of the water system; advise on issues relating to operation of the water system e.g. raising tariffs.

The WC requires the VG to: meet monthly and keep minutes of those meetings; prepare financial reports; operate the scheme daily; collect revenue by selling water; pay salaries to a village pump attendant (VPA), water sellers (WS), and security guards; liaise with the district water engineer (DWE) e.g. on a major breakdown; pay the Pump and Engine Maintenance Scheme (PEMS) contract fee.

In this management arrangement, the DWE also provides external monitoring of the scheme, as well as policy advice related to the water fund. The PEMS was established by the state to handle O&M. Villages enter into a contract with the District Water Department and pay between US$1.60 and US$4.90 per month. A technician then makes regular monthly visits to the scheme to inspect and sell parts from the PEMS shop located at the district centre. The PEMS technician is responsible for training the VPA and giving advice to the WC. The VPA is a position elected by the village and is in most cases a man. He operates and maintains the pump and engine installation, and reports any problems to the WC. The WS is also a position elected by the village, and is in most cases a female member of the WC. Four are elected, one for each tap-stand, and they earn US$0.81 for every day they sell water. They remit the daily collection to the WC treasurer at the end of each day. Presently, water is sold at US$0.03 per 20 litre bucket, available daily from 6-10 a.m. and 4-6 p.m. This revenue is used to cover O&M expenses of the scheme.

This initiative is the result of collaboration between WaterAid, a British charity, and the regional and district authorities of the government of Tanzania. The programme has been running since 1991, following changes to the Tanzanian water policy that placed the responsibility for management of rural water supplies firmly with the communities.

Source: Kashililah (1998)

142

Box 41: Village government and the water point committee: Zimbabwe

Zimbabwe is divided into five administrative Provinces, each subdivided into Districts, with each District having an elected Rural District Council (RDC) comprising of Councillors from individual Wards. Below Ward level, however, there are important differences in scale and administration when compared to Tanzania:

Households per	Ward	Village	Sub-village	Kraal
Tanzania		350-450	90-100	n/a
Zimbabwe	800-1200	150-250	n/a	20-50

In Communal Areas of Zimbabwe, the kraal is generally acknowledged as the unit with which rural people are most comfortable. Each kraal is a collection of households under the traditional authority of a kraal-head or *Sabhuku*. The *Sabhuku* is the lowest level of a traditional hierarchy, with a pyramid of Headmen (*Ishe*), Sub-Chiefs, Chiefs and Paramount Chiefs above him. In Resettlement Areas bought by the government from the commercial sector to resettle targeted sections of the population, administrative arrangements are similar to Communal Areas but without the traditional hierarchy (although there is a move to re-instate the role of *Sabhuku*).

While the kraal is universally recognized, it is not an administrative unit. The smallest administrative unit recognized by state planning authorities is the Village, administered by a Village Development Committee (VIDCO) consisting of elected representatives from several kraals. The overlapping and sometimes contradictory powers of the traditional and state systems has been a perennial subject for discussion. In Zimbabwe, user communities linked to productive water points – and their respective water point committees – invariably cut across these traditional and state administrative boundaries, and this can be a source of conflict.

The farmers, working together with their technical advisers, should thoroughly analyse their situations and identify the training requirements of them all. It is recommended that this precedes any training programme or extension activity. Training methodology should recognize that adults learn by doing. Critical analysis of their own situation is in itself a powerful learning experience for the community, and often leads automatically to solutions.

Although each scheme will identify different needs, many will focus on the same subjects, and use can be made of core training materials. The training should be practical, on-site and on-the-job, rather than classroom oriented. At some schemes there will be need to strengthen organizational capacity within the local community structures; at others to strengthen their capacity to deal with other relevant organizations in the public and private sectors.

Table 7.1 outlines potential organizational and institutional development training areas.

The training should be in the local languages, appropriate for different age, gender and class groups and the prevailing levels of literacy and numeracy. It should use practical, scheme-based examples presented in well-thought-out parables, and aim to develop manuals and plans that can be used directly by the water point committees. Examples include manuals for O&M, accounting and budgeting procedures and formats, marketing, and standard contracts (for buyers, input suppliers, service providers etc.). Continuous evaluation by trainees of training content and presentation is important to ensure that the process keeps on track.

As a rule of thumb, at least 20 per cent of scheme members, including all relevant committee members, should participate in the training sessions. Since committee members will leave from time to time, be removed from their posts, or change following elections, specific follow-up and aftercare training for new committee members will be required. It is important to ensure that members of disadvantaged groups (e.g. female heads of household, elderly women, landless families) also participate in these sessions (Box 36). To facilitate women's attendance, training should be done during school holidays

Table 7.1   Organizational and institutional development training areas

Organizational Development	Institutional Development
Community management structures and their roles	Building external relations
village government	networking skills
sub-committees	civic education
water point committee	negotiating skills
pump attendants	procurement and marketing strategies
	strategic planning
Strengthening group cohesiveness	
water point constitution	Formalizing external relationships
members' rights	social contract with the district council
participatory decision-making procedures	constitution ratified by the district council
accountability of leaders	registration of the water point by the district council
transparency in financial management	
community's values and vision	registration of the garden plot by the district council
Developing organizational capacity	bye-laws ratified by the district council
setting development goals and objectives	security of tenure guaranteed by the district council
planning and budgeting skills including tariffs	arbitration mechanisms headed by the district council
leadership / management skills	
monitoring and evaluation mechanisms	contract for pump repair with private or public sector
record keeping and accounting procedures	
financial management skills	contract with buyers
community conflict resolution skills and mechanisms	contract with extension agents

or quiet farming times and as near to the households as possible so that they are not totally disengaged from their daily activities. This will allay men's fears of women abdicating their responsibilities by going on training courses off-farm, and will help to involve men and women in the training process simultaneously.

Bye-laws

The bye-laws or rules made by the local authority cover a much broader area than a specific water point, and the process of rectifying existing bye-laws or drafting new bye-laws can take a long time. The process may go beyond the normal 'project' cycle. Nonetheless, it is an important part of institutional development.

The relevant bye-laws will generally relate to management and protection of the water points and catchment areas. For instance, user fees for O&M and participation in catchment area protection should be based on local bye-laws. Such bye-laws should be approved/ratified by the local authority to give them the force of law, and because the local authority is the ultimate arbitrator in conflicts that may arise over resource allocation. Where traditional 'bye-laws' and practices exist in relation to natural resource management and protection, these should be taken into account.

Monitoring and evaluation mechanisms

Performance evaluations repeatedly show a strong correlation between regular and transparent monitoring and overall performance outcomes. Monitoring should therefore become an integral part of the scheme improvement process, both for and by the water point committee and the external institutions involved. Two types of monitoring will generally be beneficial:

- O Programme monitoring: mainly a water point committee activity in which the members involved periodically review the functioning of the committee and productive water point, including record keeping and accounting procedures, and formulate points for action;
- O Impact evaluation: mainly a local authority or NGO activity on selected water points aimed at evaluating economic impact, water point committee effectiveness, and overall impact on the different categories of users. The results of these studies should lead to the formulation of follow-up training and adjustments in the project approach.

Management of community gardens

Determination of garden membership

The preferred criteria suggested by communities in Zimbabwe to decide membership of gardens at productive water points are shown in Table 7.2. There were marked differences between sites. At some, respondents preferred

Table 7.2 Determination of garden membership

Preferred criteria	Percentage of households in:					
	Muz'	Gok'	Dek'	Nem'	Maw'	Mat'
Kraal-heads to decide	37	73	40	23	-	7
VIDCO chairman to decide	27	33	-	33	-	-
Extension worker to decide	3	-	63	53	-	7
Garden committee to decide	24	7	-	-	-	-
The 'community' to decide	27	7	7	10	3	3
Payment of a joining fee	-	-	-	-	77	50
All interested join	-	-	-	-	17	20
Labour contribution qualifies	-	-	-	-	13	-

that village leaders should decide who should join: the kraal-heads were the most commonly cited people for this task. At others, it was advocated that agricultural extension workers should have the greatest prominence. The VIDCO chairmen were expected to play a part at certain sites, but at one (Gokota) this subsequently led to conflict with the traditional leaders. A garden committee was deemed to be relatively unimportant in taking these decisions, as one might expect, since such a body is normally chosen by the members rather than vice versa. However, at Muzondidya it was noticeable that a committee had been appointed well before membership had been decided and even before the digging of the well had commenced. A minority of respondents at each site advocated a more consensus-based approach whereby 'the community' should decide who the most suitable members would be. Payment of a joining fee was emphasized as the main criterion at certain sites. This has the advantage of avoiding any discontent arising from charges of favouritism where others make the decision, but may exclude the poorest members of society. Perhaps to overcome this danger, minorities of respondents at these sites also advocated membership open to all who were interested. A compromise to admit families contributing labour during construction, although seeming practical, was favoured by only 13 per cent of respondents prior to scheme construction. Ten per cent believed that 'the needy' (defined as widows and those families without a regular wage earner) should take priority.

In the event, 67 per cent of all members across the six sites actually joined the gardens through a mixture of contributing labour for well and garden construction and paying a reduced joining fee. Thirty-two per cent paid a joining fee alone. The mean payment to join a scheme was US$1.40 in 1998 prices.

The method of supplying labour to join a scheme has an advantage in that it promotes a sense of ownership. However, it has to be recognized that availability of labour is a dynamic variable. Labour availability is greatest during the dry season and at its most scarce during the wet season when people are busy preparing their rainfed fields. Hence the timing of project construction within the farming calendar is critical for obtaining labour and promoting ownership in this manner. There is also a danger with this approach that households with less male labour will miss out on becoming members. If these households are poor, they will not have funds either to pay a cash joining fee or to hire labour. It is possible to investigate allocating spaces in the garden in an auction systems, with some kind of equity constraint (Box 36) or to create a mortgage system for the poorest and those with no male labour to allow these members to pay back a joining fee over time with the aim of eventually buying the plot from the committee. This issue needs careful thought and open dialogue between the community and project staff during the planning phase (Chapter 6).

Membership and garden performance

Figure 7.1 shows garden membership recorded over time. It is remaining fairly static. On a positive note, this means that long after any 'honeymoon period', the benefits of membership still outweigh costs and, importantly, that local élites have not used their influence to undermine the rights of the initial members. On the negative side, static membership means relatively little scope for new households to join. Old age followed by distance to the water point were cited as the main reasons deterring those who did not wish to join at the time of garden construction, but the number of non-member households now wanting to join ranges from 10–22 per site. This highlights again the importance of involving at the outset *all* who might subsequently use the water point, by developing enabling strategies through project constitutions and other mechanisms that promote equity and avoid closed or exclusive membership (Box 36).

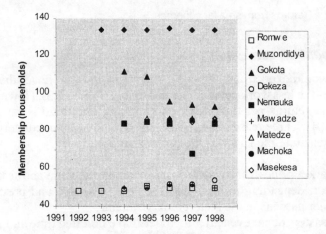

Figure 7.1   Garden membership over time

Smaller schemes in terms of membership tend to be easier for communities to manage, and tend to produce higher returns (Figure 7.2). However, the trend is not significant and the wide scatter in the data indicates that there are many other factors as well as size of membership which affect scheme performance. The worst performing scheme in one year can be the best performing scheme in the next year – these schemes do exhibit their ups and downs and this should be anticipated. The advantages of promoting larger membership, at least for the initial productive activity, include more equitable distribution of benefits within the community and more households able to participate in establishing revolving funds and additional new income-generating activities.

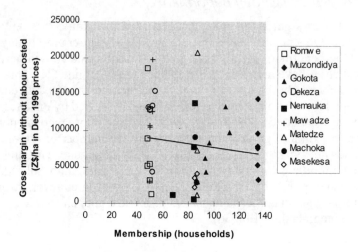

Figure 7.2   Returns from smaller schemes

Time spent on gardening

As one would expect, the higher the membership – the fewer the beds per member – the less work per member household (Figure 7.3). However, when looking at returns per time spent on gardening the trend is far less clear (Figure 7.4). Three reasons why gross margins do not immediately increase with increasing time spent on the gardens are:

O   market prices vary widely from scheme to scheme and from year to year, depending primarily on the type of season and prevalence of other gardens in the area selling vegetables;

O   disasters occur even after members have invested time and effort; for example, where livestock break into a garden or where an outbreak of pests or disease destroys a crop;

148

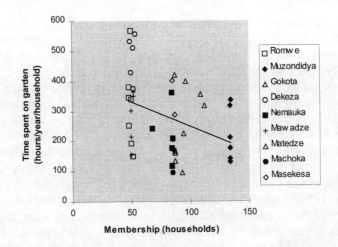

Figure 7.3   Membership and time spent on gardening

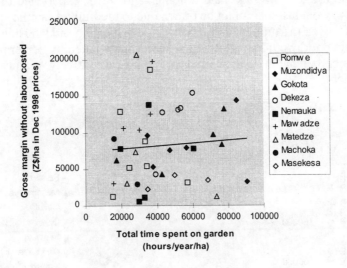

Figure 7.4   Garden performance and total time spent

○ community records tend to under-report yields; for example, where there is a break in record-keeping or where families informally take a few vegetables each night for home consumption. They also tend to over-report the time spent on the gardens, especially where approximate figures in hours/week/member given by a few individual members are scaled up to annual figures for the whole membership.

Nonetheless, traditional irrigated horticulture is a highly labour-intensive activity and time-saving interventions will be important in the future, as will increased attention to marketing strategies, pest and disease control and garden security. The importance of marketing is reflected in the positive correlation between garden gross margins and length of selling period (Figure 7.5). Gross margins are highest where garden members are actively marketing produce for longer periods of the year. A widely held belief that community gardening in southern Africa is a winter or dry-season-only occupation is a fallacy. Several schemes now actively sell at least some produce throughout the year.

Many other factors that might also be expected to influence garden performance, and which were measured in baseline and return-to-household surveys before and after scheme implementation, were found not to influence garden gross margins significantly. In a simple regression analysis the 'percentage of members classified as wealthy', 'selling period' and 'cattle ownership' were found to be positively related, 'size of membership' negatively related, with tenuous links also suggested between 'use of the water point for domestic water' (gross margin decreasing with increasing domestic usage), 'average land holding' and 'prior experience of group projects'. Conclusions cannot be drawn due to the limited sample size, but the link to wealth (implicit also in both cattle ownership and land holding) is interesting and would suggest that schemes may generally perform better where membership includes at least some who are relatively wealthy within the society.

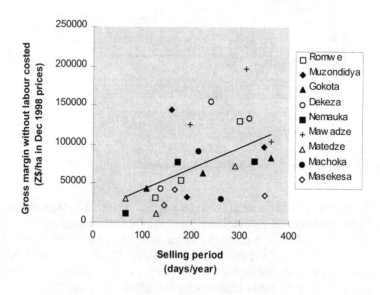

Figure 7.5 Garden performance and selling period

150

Although all-year-round vegetable production is possible, the peak season of activity for community gardens in southern Africa is the dry winter season. In summer, the opportunity cost of labour rises as fields need to be prepared for the staple rainfed crops. Farmers also cultivate indigenous vegetables as an intercrop among the staple food crops; the incidence of vegetable pests and diseases in gardens is higher; and, in good rainfall years, some low-lying gardens become waterlogged.

Surveys across six schemes show that during the winter period the amount of time per household spent working in the gardens ranges from 7.2 to 14.2 hours per week, with an average of 11 hours per week. Eighty percent of respondents say the main decision-makers for the gardens are women, who also provide most of the labour with the help of children. The task that takes the longest in all cases is said to be watering – pumping from the well and carrying the water by buckets to their beds. If time efficiency can be improved, summer vegetable crops will become more common, because the main constraint to gardening in summer is said to be shortage of labour.

In terms of time spent on the gardens over the years, Figure 7.6 shows there is no distinct trend. Some schemes consistently invest more time than others, and increasing experience of garden production and income generation does not appear to alter this. If there is any trend it probably reflects external factors such as the previous year's rainfall. This tends to influence when many members complete harvesting of rainfed crops and can begin gardening in earnest. It also determines market competition from surface-based schemes, and the importance placed on the productive water point as part of the household's risk-coping strategies in that particular year.

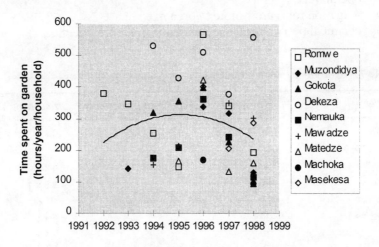

Figure 7.6 Time invested in the gardens over the years

151

Garden design and operation

Community gardens are generally divided into plots or beds, with each household privately owning a certain number or allotment. The total number of beds is usually divided equally between households, but where water supply allows there is scope for some to rent additional beds (Box 36). The normal layout of beds for each member household is in a strip running across the garden rather than as a square block. This helps ensure that distances walked are similar for all families where the water point is centrally located in the garden and it enables blocks of land to be planted with one type of vegetable to facilitate chemical pest control. Table 7.3 shows number of beds, size of bed, and layout of holdings recorded at gardens in Zimbabwe. Path widths between beds are usually less than 0.5 m.

Although there are subtle variations, management of community gardens is usually part co-operative and part individual. Where the garden forms the first project at the water point, an elected water point committee, sometimes in conjunction with the local agricultural extension officer, usually decides on such matters as the crops to be grown, the area to be grown under each crop, the timing of planting and transplanting, irrigation schedules, chemical pest control and the timing and application of pesticides. The committee is also responsible for purchasing seeds and chemicals and establishing nurseries. Individual members tend their own beds, harvest vegetables for home consumption and retain the proceeds from vegetable sales, less any payments required by the committee. This management procedure results in gardens that have a regimented appearance. The advantages of this approach are:

○ savings can be made by buying and using seeds and chemicals co-operatively;
○ irrigation rotas are simple to organize;
○ chemical pest control is made easier and more effective.

Table 7.3 Garden layout at productive water points in Zimbabwe

Garden	No. of members	No. of beds per member	Size of beds (m)
Romwe	48	7	6 x 1
Muzondidya	124	3	4 x 1
Gokota	130	4	4 x 1
Dekeza	56	16	3.5 x 1
Nemauka	74	3	12 x 1
Mawadze	30	5	8 x 1
	10	5	6 x 1
	10	10	8 x 1 and 6 x 1
Matedze	87	7	3 x 1

The disadvantages of this approach are:

O   individual choice is not catered for and self-reliance is not encouraged;
O   non-chemical pest control is made difficult;
O   a relatively large quantity of vegetables comes on to the local market at the same time;
O   all members suffer if there is a problem with, say, the communal nursery.

Garden location

Where land conditions allow, the productive water point may be located within the garden fence. This should be a community-based decision. Advantages of locating it at a central point inside the garden include reduced distance to carry water for irrigation and increased protection from animals and vandalism. Children are sometimes prone to putting stones down pump outlets! Disadvantages of siting the water point within the garden are inconvenience for people requiring only domestic water, especially if the garden is locked for parts of the day, and increased risk of theft if the garden is left open to allow access for domestic water. The NGO and supporting agencies can provide guidance. Where possible, the garden should always be sited below the water point to allow irrigation by gravity feed and to reduce the drudgery of water distribution. However, be wary of land prone to waterlogging in very wet years.

Fencing

A secure fence or wall is required to provide a sanitary space and to keep animals out. In some countries, this fence may be cut from brushwood or thorn trees but where this is considered environmentally unsound, a 'live fence', hedge or steel fence will be needed.

Live fences are closely planted, pole-like tree species. A simple live fence can be established at the start of the wet season by stringing wire along a line of truncheons (lengths of branches planted as cuttings). Once established, an effective barrier can be made by cutting branches of the fence poles half way through, bowing them down and bending and weaving the living ends into the foliage to form a fence. A live fence forms a useful windbreak protecting crops from drying winds for an area 20 to 30 times its height on the leeward side and about 4 times its height on the windward side. Shallow-rooted species (particularly exotics such as eucalyptus) should not be used around cropping areas. Live fences are also particularly useful for setting up a paddock system, and fodder species may be used where animals are kept.

Hedges are barriers of growing plants. It is often desirable to combine different unpalatable and impenetrable species. For example, truncheons of narrow spiny trees (*Commiphora africana*) can be interplanted with a combination of climbing species (*Acacia ataxacantha* and *Toddalia asiatica*) and bushy species (*Acacia erubescens* and *Carissa bispinola*). Other plants with which to grow a hedgerow in dryland areas include species of the families *Euphorbia, Prosopis, Leucaena* and sisal (*Agave americana*). A hedge requires effective management. The aim is to produce a wall of vegetation from top to bottom and side to side. It also needs to be shaped. Planting may be beneath a traditional brushwood fence. As this rots away, it is replaced by the live fence growing through it. From the time the hedge reaches a height of about 0.5 m there should be regular trimming to form an inverted V shape along the length. This maximizes the surface area exposed to the sun and reduces competition for space and light with adjacent crops. Trimming also provides mulch for the garden or fodder for livestock. Competition for water and nutrients is reduced by digging a trench up to 0.7 m deep between the hedge and the adjacent crops.

Steel Fences: as a guide, the materials required for a square steel fence of side 100 m are:

17×2.4 m $\times 48$ mm posts
10×2.4 m $\times 38$ mm stays
48×2.4 m standards
1×2.1 m $\times 1.8$ m $\times 38$ mm gate
400 m $\times 1.2$ m $\times 75$ mm $\times 2.5$ mm diamond mesh
1×50 kg roll of barbed wire
1×50 kg roll of 13 gauge straining wire
1×50 kg roll of 14 gauge tying wire
22 pockets of cement

Fencing is an expensive item, both in terms of materials and community labour. Irrespective of fence type, it is worth remembering that:

○ options for garden design should always be decided by the community in open dialogue;
○ for the same length of fencing materials, a round garden provides 27 per cent greater area than a square garden and about 40 per cent greater area than a rectangular garden;
○ fencing small gardens is relatively expensive, and for a relatively small additional sum, a larger area can be fenced, providing room for expanding membership and activities;
○ a clever gate, weighted to close on its own, can help to exclude livestock;
○ providing communities with training, materials and wire fence-making tools is a more cost-effective approach and promotes local ownership.

Pest and disease control

Baseline surveys prior to scheme implementation suggested that priorities for problem solving at community gardens would include pest and disease

Table 7.4 Main problems encountered in previous community gardens

Problem	Per cent of households in:					
	Muz'	Gok'	Dek'	Nem'	Maw'	Mat'
Pests and diseases	73	93	90	90	90	57
Water shortage	40	90	57	63	33	93
Theft	47	33	43	30	50	23
Damage by livestock	-	-	3	10	17	20

control, water saving methods of irrigation and secure fencing (Table 7.4). Other problems anticipated by less than ten per cent of respondents were crop marketing and shortages of seed, fertilizer and implements.

In the event, operational data have supported some of the baseline observations. Pest and disease control, in particular, is seen as *the* major problem at all schemes, particularly during summer months. Many vegetables are sensitive to attack, and severe attacks often result in complete crop failure. This is one of the reasons why garden members grow only a limited range of vegetables such as covo or rape in the *Brassica* family, which are relatively hardy with regard to pests and diseases. It also explains why members use hardly any inorganic fertilizer unless this can be obtained cheaply or as part of a drought-relief handout. Inorganic fertilizers represent an increase in cost of production and another investment that may be lost if the crops fail through pests and diseases. Members at all schemes prefer to avoid this risk at present and use organic fertilizers when available.

The need for good pest and disease control is understood in theory. Chemical sprays are used frequently where garden members can obtain chemicals and have access to a sprayer, and at most schemes members soon decided to purchase a communal knapsack sprayer using income from vegetable sales. However, advice is frequently requested regarding appropriate choice of pesticide, application rate and timing with respect to harvesting; most farmers are not aware of the importance of fungicides, and no use is being made of natural pest and disease controls. Table 7.5 lists the problems most frequently identified and highlights the range of natural control measures that are potentially available in addition to chemical controls. Further detail on pest and disease control is beyond the scope of this book but there are some excellent publications available (listed under Further Reading) which ideally need to be translated into village-level training materials and pamphlets in local languages.

Irrigation and water use efficiency

Water use efficiency in community gardens is generally far better than efficiency on larger, formal irrigation schemes (Figure 3.1). Productivity per unit of water increases substantially as plot size decreases because people who have to lift water laboriously from a well or dam and carry it to their

Table 7.5 Common pests and diseases in community gardens and their control (from Elwell and Maas, 1995; CIIFAD, 1995)

Pest/Disease	Chemical control	Natural control
Diamondback moth on leaf vegetables	Difficult with cabbages since it is hard to spray the caterpillar once it gets in the cabbage head. It is easier to spray rape and tsunga. Apply Malathion 25% WP at the rate of 1.2 kg in 200 litres of water per ha. Ensure the spray gets to the underside of the leaves where the larvae are.	Overhead irrigation can help wash off young caterpillars. Plant mustard as a decoy or trap crop. Keep hens in garden. Practice rotations: 6 week breaks with no crucifers grown. Dusting or spraying with agricultural lime. Use biological insecticides such as Biobit, Thuricade and Dipel. These products may not be as effective as several synthetic insecticides but help conserve beneficial natural predators and parasites which attack the diamondback moth. Plant younger cops into the direction of the prevailing wind to make it harder for moths to fly to the new plantings. Grow seedlings in isolated areas or screenhouses. Grow susceptible crops in blocks between other non-cabbage crops, with windbreaks between plots to form a barrier to the moths and a home for the natural predators and parasites.
Aphids on leaf vegetables	Aphids rob nutrients from plants by sucking sap from the leaves and growing tips. They also transmit viruses that cause leaf curling, leaving vegetables severely stunted and unsuitable for consumption.	Grow healthy plants using compost. Avoid fertilizers, particularly nitrates which produce sappy plants attractive to aphids. Increase plant diversity on the land by mixed, inter- and strip-cropping and by trap-cropping and barriers. Companion plants of garlic, chives, marigolds, parsley, onions and many herbs repel aphids. Milkweed and sowthistle are good trap or decoy plants to attract aphids away from the main crop. Sprays made up of soapy solutions or blackjack, marigold, tephrosia, clay and lime or 1 shovel full of dry cow manure mixed with 10 litres water stirred once daily for 14 days, then diluted 3-5 times before spraying.
Bagrada bug (Shield bug) on leaf vegetables	The bugs suck plant sap from the leaves and growing tips, and secrete a toxic saliva that causes death of seedlings and even whole plants.	Aromatic or strongly smelling plants like gums, lantana, khaki weed, tomato, garlic, chilli, onion or any herb. Dry the plant material, grind to a powder, mix with boiling water and allow to cool before spraying. Spray with mixture made up of a handful of the pest insects crushed and mixed into 10 litres of water. Add a little soap.
Red spider mite on tomatoes	Overuse of chemicals has built up a large resistant population while effectively removing most of the predators of the red spider mite. Check that spraying is done as infrequently as possible. Spot spray to avoid killing off predators.	Rotations are particularly important in the control of this pest. Allow natural predators such as ladybirds, mantises, lacewings and ants to be restored by avoiding use of agricultural chemicals. Maintain the humidity as high as possible by altering the spacing distance between plants, mulching, frequent irrigation. Close spacing discourages red spider mite. Increase plant diversity - a hedge of perennial pigeon pea around the garden is recommended as a breeding site for other mites which predate on red spider mite. Interplant with garlic, basil, onions and other aromatic plants.
Early and late blights on tomatoes	Use fungicide at 7-10 day intervals depending on damp or dry weather. Rainfall and overhead irrigation wash the fungicides off so reapply as necessary. Knapsack sprayers must be in good condition, fitted with a hollow cone nozzle, and the user must walk slowly to cover whole plant without the spray running off the plant.	Diseases are caused by fungi. Avoid overhead irrigation: wet the soil not the plants. Mulch the crop with grass to prevent rain or irrigation splash (and conserve moisture). Use disease-free certified seed and resistant cultivars. Scout to destroy suspect seedlings. Destroy crop residue as soon as the crop is finished, making compost but not for use on tomatoes. Increase organic matter in the soil using manure and maize stover. This will increase fertility and decrease nematodes. Rotate crops – no tomatoes on the same land for 2-3 years. Slash land around gardens to remove broad-leaved weeds and allow grasses, which are not hosts for early blight, to establish. Plant windbreaks of fodder grasses such as Napier or fruit trees such as mango, fig, banana.

beds take care not to lose any, and distribution losses, significant in larger irrigation systems, become negligible in traditionally-irrigated horticulture. However, water balance experiments show that over 50 per cent of the water they apply as surface irrigation in traditionally-irrigated horticulture is lost as soil evaporation (Figure 7.7). Furthermore, the rigid irrigation schedules they follow take little account of crop water requirements or the moisture retention properties of the soil (Box 42). These results give an indication of the potential savings still possible by adopting irrigation methods and scheduling that reduce soil evaporation at the same time as minimizing losses due to drainage and canopy interception.

The search for irrigation methods that use water efficiently has led to a wide range of technical innovations, a few particularly suited to small-scale irrigation and community gardens. The comparative advantages and disadvantages of some of these techniques are summarized in Table 7.6 and Box 43. As water and labour availability are the principal constraints in determining the size of garden that can be irrigated, there are obvious benefits to be gained from using water as efficiently as possible. It may be expected that these irrigation methods will play an important role in community gardens. However, their adoption to date has been poor and it is important to consider why.

Despite water shortage being high on the list of problems anticipated by communities, experience shows that uptake of improved irrigation technologies is affected by a complex mix of institutional, social, economic and physical factors. Water in communally-managed areas is a common property resource. This, combined with a poor local understanding of groundwater behaviour, immediately places limitations on the responsibility felt by individuals to conserve the resource. Important questions also relate to tenure within the gardens, to institutional management arrangements, and to why resource-poor farmers might want to adopt improved irrigation technologies: to save time? to improve yields? to save water? for themselves or for others? Essentially, water saving depends on incentive and a number of factors can work against this in community gardens. Merely introducing members to water saving technologies will not, on its own, create this incentive. Other conditions will generally need to be in place first, such as:

O   some form of pricing or regulation over water distribution;
O   open rather than closed membership;
O   access to information;
O   demonstrable time saving;
O   a general water shortage.

At its worst, community gardening may see members running hurriedly between their beds and a central tank in an effort to take water before the tank empties and to maximize their share of the common property resource. In contrast, community pricing of the water, or at least some form of

Box 42: Irrigation scheduling in community gardens

Community irrigation schedules in Zimbabwe were compared with FAO recommendations. All communities practised basin or flood irrigation using water carried in buckets. About 20% of gardeners interviewed determined when to irrigate by looking at the soil – if it looked dry, they irrigated. About 12% irrigated when they observed signs of crop wilting. The rest tried to make it a routine to irrigate twice or three times per week unless there was rain. All believed that the young plants needed to be watered more frequently to ensure good establishment. The amount of water applied depended more on proximity to the water source than on crop type. Tomatoes generally received more water than rape or cabbage, with maize receiving less. For all gardeners and all crops, more water was applied during the first part of the season than recommended by FAO. For all gardeners and all crops, except maize in some cases, the FAO figures were also exceeded for the remainder of the season. The scheduling patterns suggest that gardeners are not being economical with their irrigation water; however, some interesting and sensible methods of economizing on water use by flood irrigation were revealed by some gardeners: planting three rows per bed so that the canopy between plants closes rapidly, reducing soil evaporation; irrigating late in the afternoon; applying the maximum amount of irrigation at one time rather than small amounts frequently; practising staggered planting. In terms of crop yield and water use efficiency, some gardeners were achieving yields comparable to a local agricultural experimental station but all were using more water, resulting in lower efficiency.

	Yield (kg ha⁻¹)		Water applied (mm)		Water use efficiency (kg m³)	
	garden	expt	garden	expt	garden	expt
tomatoes	29860	52700	668	487	4.47	10.8
cabbage	28353	77000	638	461	4.44	16.7
maize	10458	13300	537	440	1.95	3.02

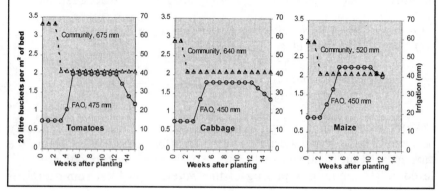

158

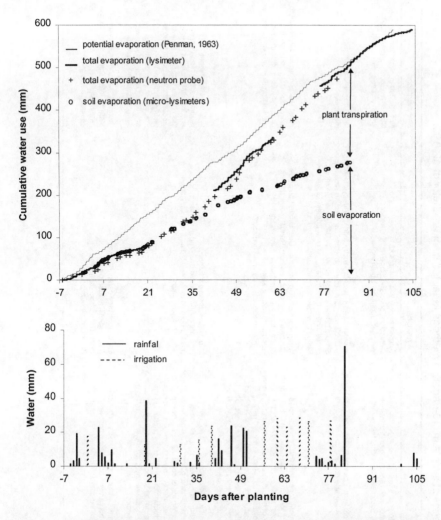

Figure 7.7 Soil evaporation and plant transpiration under flood-irrigated maize (source: Batchelor et al., 1996)

partitioning among the users, will promote uptake of methods that make more effective use of the allocation available. Similarly, closed or exclusive garden membership will tend to result in 'members' feeling relatively secure in the knowledge that they (generally) have sufficient water for their own needs, with little concern or incentive to save water for others, whereas open membership and increasing numbers of users creates pressure on individuals to make more effective use of the limited allocation available.

Communities, and water point committees in particular, need access to information on aspects such as the sustainable yield of the water point and corresponding irrigable area, the status of the groundwater resource under

Table 7.6 Advantages and disadvantages of simple micro-irrigation methods (adapted from Batchelor et al., 1996)

Method	Description	Advantages	Disadvantages
Traditional flood irrigation	Involves carrying water in buckets to pour it over raised vegetable beds, typically 5–10 m long x 1 m wide, often with raised edges to prevent runoff. The surface of the beds is generally bare apart from one, two or three rows of crops along the length.	Traditional, well known method. Easy to perform. Good crop establishment. Minimum additional inputs required.	No efficient in water use. No inherent control against over-irrigation. Labour intensive.
Low-head drip irrigation with dripper strings	Designed to irrigate 150-200 m² circle from a central 200 litre drum set on blocks 1m high and connected to home-made drip lines.	Improved water use efficiency. Good uniformity of wetting. Suitable for saline/ brackish water. No filtration required. Simple to make, easy to use. Low labour requirement. Some inherent control against over-irrigation. Durable. Suited to small plots and sloping land.	Initial cost and availability of materials.
Subsurface pipe irrigation	Involves burying home-made clay pipes (or slotted PVC or perforated bamboo) along the centre of 5–10 m long vegetable beds with 0.1–0.2 m of soil above the pipes. The clay pipes are 0.25 m long, 0.075 ID and 0.115 OD fired in a shallow pit but not glazed. An inlet is formed at one end by tilting the first pipe section, the lower end of which is angled during manufacture. The other end of the pipe is blocked. To irrigate, water (and fertilizer in solution) is poured into the inlet using a bucket or hosepipe. Water from the subsurface pipe seeps directly into the root zone via the joints between the individual pipe sections or perforations/slots.	Improved water use efficiency. Clay or bamboo pipes can be made locally. Low cost. Robust method. Some inherent control against over-irrigation. Good uniformity of wetting. Simple to use. Inherent weed control.	Initial skill and labour requirement for manufacture and installation. Crop establishment can be poor if initial irrigation only via pipes.
Pitcher irrigation	An ancient method of micro-irrigation in which porous clay pots or unglazed earthenware pitchers are buried in the soil with just the neck above the surface, either adjacent to individual plants, at the centre of a circle of plants, or between plant rows at intervals of about 0.3 m. Pitchers of OD about 0.2 m and capacity about 2 litres, necks about 0.1 m long, are most practical. When filled, water seeps through the walls of the pitchers to the soil creating localized wetting zones similar to those produced by subsurface drip sources. The amount of water applied at any one time is controlled by the volume of the pitcher and frequency of filling.	Improved water use efficiency. Inherent control against over-irrigation. Can be positioned next to individual plants or in very small plots or sloping land.	Initial skill and labour requirement for manufacture and installation. Less robust than subsurface pipe. More labour-intensive, pots have to be filled individually. Porosity of pots decreases with time. Difficult to cope with high crop water requirements.
Flood under mulch	As for traditional flood irrigation, but the water applied directly to the soil surface beneath a layer of mulch. This can comprise native grasses, maize stover, straw, leaf litter, coarse sand and even flat stones arranged across the surface between plants.	Easy. Good crop establishment. Adds organic matter to soil. Protects fruit from damp soil. Prevents soil capping.	Potential for increase in pests. Material for mulching not always readily available. No control against over-irrigation,

160

Box 43: Wagon-wheel low-head drip

Materials:
6 × 7 m × 15 mm polythene class 3 pipes (driplines)
7 × 15 mm nylon T-pieces (manifold)
1 × 100 mm × 15 mm nylon running nipple with 2 nuts
1 × 15 mm tap (to drum)
5 × 15 mm × 0.3 m polythene class 3 pipes (manifold)
1 × 200 litre drum cut open at one end
1 × roll nylon braided string (drippers)

Set the tap into a hole drilled near the base of the drum. To this attach the circular manifold comprising T-pieces and short lengths of pipe. Fit driplines to the T-pieces and lay outward as spokes of the wheel. Mark the position of the holes on the pipe 0.3 m apart, heat a 1.5 mm diameter nail and burn a hole through the pipe. Cut 0.1 m pieces of string and thread through the two holes. Tie a knot on each side. Close the dripline at the end by bending the pipe and tie it with a piece of wire. A well planned wheel system has the potential to produce 500 kg of vegetables annually. Irrigation is 200 litres of water three times/week. Plant vegetables along the driplines next to the wetting area. Clean clogged drippers by pulling the string from side to side. Each member can own a complete wheel system or individual driplines where membership is high. An improved micro-environment is created by adding extra T pieces and driplines to form the rim of the wheel around which a windbreak crop of maize, grapevines or other tall crop can be irrigated.

(Source: Albertse, 1995)

present and proposed use, appropriate irrigation schedules for the different crops grown, appropriate micro-irrigation systems based on tenure, holding size, soil type, crop type, managerial skills and economic status. Opportunities for local communities to see new ideas through exposure visits rather than just being told about methods will be important. Extension staff should encourage the farmers to carry out their own experiments based on what they have seen, and it is important that extension staff and farmers are willing to learn together. Lastly, interest and investment in water saving technology is likely only when water is relatively scarce, and is unlikely during a wet cycle (Figure 2.5).

At productive water points in Zimbabwe, reducing the time taken to complete the irrigation is generally seen as more important than saving water (although for the present systems of irrigation, water-efficient irrigation would equate to time-efficient irrigation). Thus members are interested first and foremost in pump technologies that reduce the drudgery and time taken to lift and distribute the relatively large volumes of water needed for irrigation, rather than in irrigation methods. Investment in improved pump technology that can lift water to an elevated tank, for example, will repay itself as time saving through irrigation by gravity feed (Chapter 3) and it is this type of intervention that is likely to be the first choice of members rather than a water saving irrigation method *per se*.

Crop marketing and diversification strategies

Market outlets for vegetable and fruit production are imperative for successful small-scale irrigated horticulture to occur. Relative proximity and reliable physical linkages to a market must exist. Without a sufficient market for vegetables and fruits, increased market garden production will not have the desired effect of increasing producer household incomes.

At productive water points in Zimbabwe, crops are marketed in three main ways: by the farmers direct to consumers in the local villages and market centres; through co-operative actions; and through contracts with buyers or middlemen from nearby towns. Table 7.7 summarizes crop marketing strategies adopted at seven schemes.

Except when farmers have firm contracts with buyers, there are big risks. It appears that the market information system for growers and traders on prices, expected volumes of output, and sources of support, is non-existent or very poor. Agricultural extension services are not trained in marketing irrigation produce and are not providing this service in Zimbabwe. Both farmers and traders operate in a vacuum. Generally, most members of productive water points market their produce individually at the gardens or at local market places within 5 km. Schemes within a 20 km radius of towns may also target the town market, and in some cases buyers from these markets will set up contracts with individual growers on a one-to-one basis to buy their produce. But generally there is poor market intelligence and farmers have poor negotiating skills. Most farmers do not monitor the weekly fluctuations in prices to keep themselves fully conversant, and most are not aware of marketing alternatives. Because of this lack of understanding of the markets, they do not take sufficient account of the expected seasonal market behaviour when planning the scheduling of crop planting. This is crucial to the timing of the harvesting and the marketing of the crop, and may result in a community failing to secure a niche in the market for their produce.

There is considerable scope to try new cropping strategies as a means of maximizing income from the gardens. These strategies include members selecting different crops (ranging from herbs and spices to flowers that will be pressed for essential oils) on an individual basis rather than by committee decision. The key factor is probably the importance that garden members attach to income generation as opposed to growing vegetables for home consumption, and this is determined largely by scheme membership and the corresponding land holding available to the members. If income generation is the highest priority, and members have sufficient beds, there is an incentive to adopt their own cropping strategies. However, this will require increased

162

organizational skills, especially where a spread of types and timing of planting is adopted, for activities such as buying of inputs, scheduling irrigation and cultural operations, and for organizing labour, transport and marketing.

In general, the potential for an expanded market will depend on five factors: increased local demand; increased contact with non-local buyers; increased diversification; improved transport; and interventions that extend the agricultural calendar, such as improved water lifting and distribution technologies and improved pest and disease control. External support agencies can play a key role in making this happen and ensuring that productive water points have a positive impact on the wider production systems. Two areas of focus include facilitating contract farming between individual growers and commercial buyers, and promoting joint ventures between several schemes and large market outlets.

Project diversification will also be important. Respondents at all schemes are now unanimous that water shortage as a perceived problem has decreased in importance. The pressing needs and priorities for the future now depend on who you talk to (Box 44). People are divided around opportunities for additional farm inputs (for those who own land), and projects that create job opportunities not necessarily related to farming and which do not depend on land. It is clear that project diversification at productive water points will be important, and that initiatives such as livestock feedlots and small-scale dairy units, which combine past experience of keeping livestock with the present need for income, will be important.

Box 44: Examples of priority projects

Older men	1) Cattle for draught power
	2) Cattle for pen-feeding (cash)
	3) Tractor and trailer to transport produce
Younger men	1) Cattle for pen-feeding (cash)
	2) Grinding mill (cash)
	3) Small-scale irrigation (cash)
Older women	1) Knitting machine (cash)
	2) Sewing machine (make own clothes)
	3) Windpump (reduce effort to pump water)
Younger women	1) Sewing machine (cash)
	2) Poultry-keeping (cash)
	3) Pre-school for the young children

Table 7.7 Crop marketing strategies at productive water points

Site	Members	% sold	Price setting	Buying / Selling Practice
Romwe	50	75% April-May when few gardens, 50% June to Sept, 25% Oct-Feb limited by vegetables available to sell although high demand.	Survey by members at local townships on a season-by-season basis	In 1992-94 when there were few gardens and vegetables were in demand contracts were initiated by buyers mainly at local townships up to 10km away. A buyer may have a contract with 3-4 individuals in the garden and even put down a deposit to ensure they would get their vegetables. A selling day each week was agreed when the buyers would visit the garden, the quantity bought determined by quantity harvested. Now the situation has reverted to a buyers' market.
Muzon-didya	134	75% in summer when families have other vegetables in rainfed fields; 50% mid-June to August	Members set the price based on demand on a season-by-season basis	Members on an individual basis to locals and visitors. Buyers include local township, teachers at two schools, non-members and even members who are short. No formalized contracts. Rarely to towns 50 km away to sell. Rarely exchange vegetables for grain. If vegetables not sold they are cooked and dried for use later in the year.
Gokota	93	75% May-July; 50% Aug-Oct	Members set the price but based on demand e.g. $6/kg when vegetables scarce, $5 when abundant.	Members on an individual basis to locals and visitors. 90 per cent are local buyers including schools and township, 10 per cent from towns up to 20 km. Members will take vegetables to these outside buyers, except in Jan-Apr when demand is high and buyers will visit the garden.
Dekeza	54	100% Feb-Apr 75% May-July 50-75% Aug-Oct when vegetables scarce	Committee and members decide based on prices for other basic commodities such as milk.	Sell individually but pay a monthly subscription to the committee. Regular buyers within 5 km include teachers at two schools and local township. Also visit town at 30 km, again selling individually. From Feb-April buyers come from up to 50 km away if produce available and up to $100/day/member can be made from sales.
Nemauka	84	75% March-May when high demand; 25% Dec-Feb when people have rainfed vegetables.	Members meet to decide on season-by-season basis e.g. double price in Sept-Oct when vegetables scarce.	Locals from two townships, a clinic, teachers from two schools, non-members and even members who are short, travel up to 5 km to the garden to buy from individual members. Some buyers come from 17 km away to buy tomatoes but not leaf vegetables.
Mawadze	50	75% Sept-Oct when demand high; 50% May-Aug when demand is from locals and outsiders.	Committee and members watch local market price, and sell at half this to give market traders their 100% profit. No discount given for bulk purchases to these traders, however. Fully recognize the market potential in summer and strive to grow even in difficult years.	Actively marketing. In Sept-Nov, individual members contract to sellers at a town market 10 km away, sometimes taking to market, buyers sometimes coming to garden. If a member fails to meet demand, others with vegetables to sell can take their place. When competition from other gardens is high June-July, members co-operatively sell in town 80 km away sending one person to do transaction but who has to accept the price offered at the market. Again in June-July members trade for either maize or groundnuts. Two members have invested in market stalls so no longer sell locally but maximize profit selling at the town.
Matedze	87	75% May-June when no competition, 50% July-Aug, balance consumed or dried for use in Oct-Dec.	Committee members set price on season-by-season basis based on visit to roadside market 2 km away.	Selling is done individually to locals within 4 km who include township, non-members and even members, rarely to buyers from town at 50 km. No formal contracts. Few customers reported in Nov-Feb.

Productive water points and the surrounding environment

Over-exploitation of groundwater

As earlier chapters have shown, current use of groundwater in Africa is generally low. Conventional rural water supply provides point sources for domestic use only, and total abstraction by this degree of development equates to less than 0.5 mm of recharge where present evidence is that actual recharge can be much higher, in the range 50-150 mm. Failure of wells and boreholes under these conditions is not due to resource use, or even to human impact on catchment hydrology through land use change, but is caused by inadequate siting and design of the water points to cope with the low rainfall periods, high natural groundwater recession, high spatial variability and low permeability that typify hard rock aquifers in dryland areas.

Elsewhere, however, most notably in the Indian subcontinent, the water balance has been changed in recent years, and failure of wells and boreholes in these places is now popularly attributed to 'over-exploitation' of the groundwater resource. Some 80 per cent of drinking water needs are now being met by groundwater, and recent policies to subsidize the costs of drilling and of water lifting pumps, combined with rural electrification, have led to massive development of boreholes and agrowells to support small-scale irrigation.

Agrowells are typically 6 m in diameter and use mechanized pumps to draw upward of 75 m^3 of water per day over a growing season of some 180 days from mid-February to September. Well spacing is reported to vary from only 50 m to 500 m (a density of one well per 10–20 ha) and this development now equates to at least 70 mm of recharge. In stark contrast to Africa, annual extraction of groundwater in these parts of the Indian subcontinent now equals or exceeds annual recharge in many years.

Major water shortages are predicted. If current trends continue over the next 50 years, it is predicted that India will no longer be able to meet the demand for water from the country's 1.57 billion people. Water availability per person has already gone down from 6000 m^3 in 1947 to 2000 m^3 in 1997 and could fall to 760 m^3 by 2047.

The development of agrowells is largely in response to increasing population pressure and the need for agricultural intensification (Box 3), but it has taken place in a haphazard, uncontrolled way without general assessment of the hydrological properties of the aquifers, the possible yield, long term behaviour, or rational siting of the wells.

Given that this scenario may apply to Africa in a few year's time, it is important to consider the concept of 'safe yield' and groundwater development strategies that will avoid over-exploitation of the resource by learning from the lessons of India and Sri Lanka.

The concept of 'safe yield'

Traditionally, 'safe yield' has been defined as the attainment and maintenance of a long-term balance between the amount of groundwater withdrawn annually and the annual amount of recharge. Thus, 'safe yield' limits groundwater pumping to the amount that is replenished naturally. Unfortunately, this concept of safe yield ignores discharge from the system. Under natural or equilibrium conditions, recharge is balanced, in the long term, by discharge from the aquifer into streams, springs, or seeps, and through transpiration by deep-rooted vegetation. Consequently, if pumping equals recharge, eventually streams, springs and seeps dry up and transpiration by deep-rooted vegetation is reduced. If pumping is in excess of recharge the aquifer is eventually depleted. This has happened in various locations across the world.

Despite being repeatedly discredited in the literature, 'safe yield' continues to be used as the basis of groundwater management policies, leading to continued groundwater depletion, stream de-watering, and loss of wetland and riparian ecosystems. To better understand why 'safe yield' is not sustainable yield, a review of hydrological principles is required. These are concisely stated by Theis (1940), Bredehoeft et al. (1982), Bredehoeft (1997) and Sophocleous (1997). Appendix 4 provides detail.

Under natural conditions, prior to development by wells, aquifers are in a state of approximate dynamic equilibrium: over hundreds of years, recharge equals discharge. Discharge from wells upsets this equilibrium by producing a loss from aquifer storage. A new state of dynamic equilibrium is reached only by an increase in recharge (induced recharge), a decrease in natural discharge, or a combination of the two. Initially, groundwater pumped from the aquifer comes from storage, intermittently it can come from reduced discharge, but ultimately it comes from induced recharge. The timing of this transition, which can take a long time by human standards, and which depends on the temporal patterns of recharge and discharge, is a key factor in developing sustainable water use policies. However, it is exceedingly difficult to distinguish between natural recharge, induced recharge and reduced discharge (from vegetation) to ascertain possible sustained yield. This is an area that needs further research.

As Sophocleous (1997) explains, science will never know all there is to know. Rather than allowing the unknown or uncertain to paralyse us, we must apply the best of what we know today and, at the same time, be flexible enough to allow for change and for what we do not yet know. Hence, in groundwater management, we should recognize that:

○ The response of a groundwater system depends on the aquifer parameters of transmissivity and storage, the aquifer geometry, and the placement of wells in the system with respect to recharge and discharge zones.

○ Although undisturbed rates of recharge or discharge are interesting to know, they are almost irrelevant in determining the sustained yield of a particular groundwater reservoir. This depends on how much the rate of recharge and or discharge can be changed and the hydrologic effects that you want to tolerate, ultimately or at a given time, and which could be dictated by environmental, social or economic factors.

○ Equilibrium is reached only when pumping is balanced by capture, defined as the decrease in discharge plus the increase in recharge (induced recharge). If the change in recharge is small, it will depend on how much of the natural discharge can safely be captured.

○ Before any natural discharge can be captured, some water must be removed from storage by pumping, hence some groundwater must be mined before the system can be brought into equilibrium.

○ In many environments, the dynamics of the groundwater system are such that it may take a long time for any kind of equilibrium condition can develop. In some environments, the system response may be so slow that mining will continue beyond any reasonable planning period.

○ Instead of determining a fixed sustainable yield, managers should recognize that yield varies over time as environmental conditions vary.

With uncontrolled groundwater development, it is inevitable that many individual water points haphazardly sited on the spatially highly variable aquifers typical of basement rock will be located in less than optimum locations. Farmers irrigating from shallow wells on the edge of the aquifers and in areas of shallow weathering may suffer irreversible setbacks if other farmers on deeper parts of the aquifer have access to submersible pump technology and deeper wells. Although it will not be the first choice of these farmers who own the better water points, improved groundwater management in these conditions will depend on collective action and legislation which promotes development of fewer but high yielding community water points at these prime groundwater locations.

Part of the current problem of groundwater 'over-exploitation' in dryland areas of India and Sri Lanka is that traditional practices of conserving and managing water, which revolved around collective wisdom and collective ethics to maintain common assets like ponds and community wells, are gone, and in some regions not a single community well has been built in the past century. Collective management of groundwater is discussed further in Chapter 8.

Rainwater harvesting to induce groundwater recharge

A second critical factor for sustainable groundwater development where population pressure is high is water conservation through rainwater harvesting to increase (induce) groundwater recharge. As we have seen in Chapter 2, natural groundwater recession through lateral flow, deep percolation and abstraction by deep-rooted vegetation, is a major process that annually accounts for a large portion of any recharge. Moreover, natural recession is continuous, whereas recharge is an intermittent process that may or may not occur in any year. The practical consequence is that recharge in a particular location may have little lasting benefit, and carry-over from year to year is limited by the natural recession. Hence, groundwater management depends on inducing recharge to ensure at least some replenishment each year.

A close relationship exists between recharge and catchment relief, surface water hydrology, soil and vegetation cover. Recharge is sensitive to land use changes, most notably increased surface runoff associated with desertification. At catchment scale, runoff from dryland areas is generally a small part of the water balance, typically less than 10 per cent of rainfall, depending on soil type. But in volume this can still be significant, and some 100 times the total groundwater abstracted by communities for domestic use and small-scale irrigation (Table 2.2). It is a part of this runoff that can potentially be harvested to enhance recharge.

Relatively substantial structures will be needed. With infrequent but high intensity storms, only numerous low-cost structures such as land terraces or dead-level contour channels (*Fanya juu*) or substantial structures such as small earth dams or 'tanks' will store sufficient runoff to enhance recharge in a meaningful way. Ironically, in Africa it is contour channels constructed previously to carry water away in a controlled manner to reduce soil erosion that today offer potential as groundwater enhancing structures. In many areas, these contour channels have not been maintained and contribute little to conserving soil (recognized now to depend more on in-field management of soil cover). Adaptation of these contour channels to hold rather than transport water away is a cost-effective measure that will immediately enhance groundwater recharge.

From the above, it should be noted that rainwater harvesting to enhance groundwater recharge is not the same as rainwater harvesting to improve crop production. In fact, surface management practices such as tied-ridge-and-furrow, which harvest rain where it falls and which can greatly benefit rainfed crops, may actually prevent groundwater recharge by limiting surface redistribution and concentration of runoff behind the larger conservation structures.

Soil type is important when deciding appropriate management practices. It is particularly important to detect any shallow impermeable horizons using a soil auger. Recharge will be 4–5 times greater, for example, on deep red soils overlying mafic pyroxene gneiss than on sandy grey duplex soils overlying leucocratic gneiss. In this example, the areas of deep red soil should be the focus for developing wells and should be managed to enhance recharge using modified contour channels (*Fanya juu*) and infiltration pits. In contrast, the areas of grey-duplex soil with an impermeable clay horizon at 1–2 m depth will contribute little to recharge but will be the major source of runoff and interflow. This could be harvested in small dams and used to rear fish, livestock and perhaps irrigate pastures. Infiltration pits will be of little value on these duplex soils, but in-field water harvesting will be appropriate, probably as a system of broad ridges and furrows across the contour that allow flexibility when growing crops in this periodically waterlogged environment. In dry years, staples such as maize can be grown in the furrows and benefit from localized rainfall concentration. In wet years, when a perched water table forms above the clay horizon, maize can be grown on the ridges and rice grown in the waterlogged furrows, the excess water from the furrows flowing to the small dams.

When rainfall is low or evenly distributed in the year, there will be a trade-off between the benefits to individual farmers of improved rainfed crop production through in-field rainwater harvesting, and the benefits to the wider community of enhanced groundwater recharge through rainwater harvesting at the larger scale. Another example of a trade-off is that between the benefits from fast-growing trees planted on the shallow aquifers, or the benefits of using this groundwater for activities such as small-scale irrigation. A 'whole catchment' approach to water resource management is needed, and projecting the outcome of various management scenarios using catchment management models will be important to achieving the optimum balance.

Pollution vulnerability of the groundwater aquifers

Concern has been expressed about the potential impact on groundwater quality from the intensification of agriculture around productive water points. This impact could be felt in relation to bacteria, nutrients or pesticides. Bacteriological risk of groundwater contamination from poorly designed water points and poorly sited latrines is detailed in Chapter 5. This section outlines the chemical risk to groundwater quality posed by irrigated community gardens.

Environmental problems associated with the intensification of agriculture have been apparent in the northern hemisphere for some 20 years or so. The link between a combination of expanding cultivated areas and increased fertilizer use and nitrate concentrations in groundwater has been extensively researched in both Europe and North America. Expensive measures to control nitrate pollution from agricultural sources are being implemented. More recently, pesticides originating from intensive cultivation have been identified in groundwater used for potable supply in Europe and North America, also requiring expensive measures to meet drinking water standards. While much of the work in temperate zones relates to cultivation without irrigation, nevertheless the same principles can be expected to apply to irrigated agriculture in dryland areas.

However, much less information exists from which the impact of agriculture on groundwater quality in dryland areas can be appraised. Recent studies indicate that for the most vulnerable aquifers intensive agriculture can produce serious deterioration in groundwater quality, resulting principally from the leaching of nitrogen fertilizers.

Awareness of the leaching of nitrate has led to concern about the possible leaching of pesticides. As for fertilizers, the increase in pesticide usage is now more rapid in some developing countries than in the developed world. There is evidence of significant pesticide occurrences in groundwater in agricultural areas of Europe and North America, but hardly any routine monitoring of pesticides is carried out in developing countries. This is because of the sophisticated and costly analytical procedures, the wide range of compounds in common use, and the care required in sampling (Chilton et al., 1995).

The risk of an aquifer becoming polluted depends on the interaction between the natural vulnerability of the aquifer and the pollution load imposed on it as a result of human activity. The pollution loading can be controlled or the source of pollution removed. The vulnerability of the aquifer cannot be changed; it is determined by the intrinsic properties of the material of the aquifer and the strata above it. An assessment of pollution vulnerability thus requires knowledge of those characteristics which determine the likelihood of the aquifer being adversely affected by an imposed contaminant load. These characteristics are:

- the thickness and nature of the strata making up the unsaturated zone;
- the physicochemical attenuation capacity of these strata, resulting from chemical reactions, adsorption and biochemical degradation.

Aquifers and the material above them can thus be assessed on the basis of their overall lithology and permeability and the depth to groundwater, and many models have been developed for mapping aquifer vulnerability. Taking extreme cases, a fractured coral limestone aquifer with very thin soil cover on a low-lying island with water table just below the ground surface would be highly vulnerable. A deep sandstone or limestone aquifer overlain by a significant thickness of confining clays would, in contrast, be relatively well protected from pollution.

In the case of hard rock aquifers, and basement aquifers in particular, although the water table is often relatively shallow, both soils and aquifer material have low permeability. The almost ubiquitous rise in water level after striking water is an indication of the semi-confined nature of the aquifer. Permeability is low, recharge is low, and unsaturated zone travel times are likely to be long, allowing adequate time for pollutant attenuation. The regolith aquifer is therefore considered to have relatively low vulnerability to pollution from agricultural activities. Exceptions could occur, however, where local recharge reaches the water table more rapidly through fractures, quartz veins or stone lines. Basaltic aquifers may also be significantly more vulnerable to pollution because of the thinner soils, different weathering patterns and greater potential for more rapid transport of pollutants to the water table through the jointed and fractured lavas.

Fertilizer usage in community gardens

Although nitrogen fertilizer usage is now increasing in some developing countries, the use of such fertilizers remains low in community gardens in southern Africa, at least. Many garden members either cannot afford to purchase chemical fertilizers, which in any case are often not available in rural stores, or prefer to avoid the additional expense while the risk of crop failure remains high through pests and diseases. Most gardeners apply animal manure when available, preferring goat manure to cattle manure and considering poultry manure to be even better. Application rates of one or two 20 litre buckets of manure to each 6 m^2 bed per crop could be equivalent to a nitrogen application of some 150–250 kg N ha^{-1}. While this appears high, it is unlikely to be applied to all or even many of the beds in each garden. During drought, for example, in the absence of cattle many garden members use composted leaves as a substitute.

Estimates can be made of possible groundwater nitrate concentrations, if the volume of recharge and the leaching losses are known. In the absence of actual data for either of these, however, such estimates must be considered as indicative only. Thus, assuming a leaching loss of even 50 kg N ha^{-1} from the fertilized garden plots and a recharge of 50 mm y^{-1} from the irrigated land, nitrate concentrations in the recharge immediately beneath the garden could be of the order of 100 mg NO_3-N l^{-1}. However, this relatively high concentration applies to only a small percentage of the total recharge to the water point. The proportion of recharge affected would be more significant where the well is within the garden boundary. Using a total annual groundwater abstraction of some 5500 m^3 (15 m^3 d^{-1}), and an overall recharge figure of some 25 mm y^{-1}, the area contributing recharge to the well is some 220 000 m^2, compared with the garden area of say 5000 m^2. Nitrate concentrations in the balance of the recharge are likely to be negligible. It is unlikely that, with present inputs, nitrate concentrations exceeding the WHO guideline value of 11.3 mg NO_3-N l^{-1} will be observed in the water points. This might not apply, however, if future development success in the gardens results in greatly increased fertilizer inputs and or greatly increased irrigated areas.

Current nitrate concentrations in groundwater drawn from productive water points were shown previously in Table 5.6. At all sites in the weathered regolith of granites and gneisses, nitrate concentrations are negligible (less than 1 mg NO_3-N l^{-1}), reflecting the low historical inputs from traditional dryland farming. At the two basalt sites, however, observed nitrate concentrations were twice the WHO guideline value, perhaps reflecting the greater vulnerability of the fractured basalt aquifer to pollution from the surface. This may come from nearby pit latrines or from direct ingress around the well; it may be more difficult to seal wells effectively in a fractured formation than in the weathered regolith.

171

Pesticide usage in community gardens

Pesticide usage in community gardens is also limited by local availability and by cost. Nevertheless, a range of compounds are in regular use, predominantly insecticides. A preliminary estimate of their susceptibility to leaching can be made from consideration of the physiochemical properties of the pesticides (Table 7.8). Together with evidence of their occurrence in groundwater from other regions, this can be used to select target compounds for monitoring, although there are few published studies for any of these compounds.

The mobility, solubility and degradation classes used in Table 7.8 are arranged such that the low numbers indicate those that are most likely to be

Table 7.8   Susceptibility to leaching of pesticides used in community gardens

Active ingredient	Use	Type	Acute oral toxicity[1]	Mobility class[2]	Solubility	Soil half-life[3]	Average application (g ha^{-1} yr^{-1})	Detection in groundwater
Malathion	I	OP	III (1400)	5	4	1/2	50	
Carbaryl	I/G	C	II (850)	4	4	3	1000	
Amitraz	H	A	III (800)	6	6	1	300	
Dimethoate	I	OP	II (300)	2	2	1/2	100	Unsaturated zone, Spain
Diazinon	I	OP	II (300)	5	5	3/4	70-120	San Joaquin valley, USA
Lindane	I	OC	II (90)	6	6	4	-	Found at very low concentrations in many areas
Benomyl	F	C	V (>10000)	5	6	4	140-550	
Mancozeb	F	Zn-TC	IV (>5000)	5	6	1/3	1.4-1.9	

[1] WHO classification 1988-89 on LD50 (mg/kg) rates but adjusted to include dermal toxicity and other factors, from *The Agrochemicals Handbook* (1993).
[2] Based on K_{oc} and K_{ow} partition coefficients. (1=most readily leached; 9=unlikely to be leached (strongly sorbed), after Briggs (1981).
[3] From *Pesticide Manual*, 8th Edition 1987, Huang et al. (1994) and Rao and Alley (1993)

Use	Type	Solubility class	Soil half-life classes
H = herbicide	A = amidine	1 = >100 g/l	5 = <10 d
I = insecticide	P = pyrethroid	2 = 10–100 g/l	4 = 10–30 d
F = fungicide	OP = organophosphorus	3 = 1–10 g/l	3 = 30–100 d
G = growth regulator	OC = organochlorine	4 = 0.1–1 g/l	2 = 100–300 d
	TC = dithiocarbamate	5 = 0.01–0.1g/l	1 = >300 d
	C = carbamate	6 = <0.01 g/l	

leached, and the high numbers those that are unlikely to penetrate beneath the soil. The compounds that are most likely to reach groundwater are those that are soluble, mobile and persistent. Thus, of the compounds listed in Table 7.8, lindane, dimethoate and diazinon have been observed elsewhere in existing monitoring programmes or have been shown to be leached to groundwater. For the compounds listed and at the application rates given in Table 7.8, it seems unlikely that significant pesticide residues could infiltrate to the water table and thence to the productive water points. A potential cause for concern, however, might be the washing of knapsack sprayers close to the well, and extension advice should be provided to guard against this.

It is recommended that a simple routine of groundwater quality monitoring be established at all productive water points (see also Chapter 5). Annual samples of groundwater should be taken for bacteriological and major ion analysis and, where facilities are available, for pesticide analysis.

Drought management

Drought preparedness

Groundwater provides a more permanent water supply and is less vulnerable to drought than surface water. It provides a natural buffer against individual dry years and does not fail immediately in the first years of poor rains. Failure of wells and boreholes in drought, can be foreseen, therefore. Long-term trends in groundwater levels reflect long-term cycles in rainfall (Figure 2.5). By monitoring groundwater levels and rainfall cycles, groundwater drought can be predicted in advance and allow sensible drought mitigation programmes to be planned, doing away with the need for hurried 'emergency drought relief projects' and their attendant problems. The monitoring of groundwater levels and rainfall cycles is of national importance, but can readily be undertaken at community level where the information is also needed for local planning.

Managing during the drought

Productive water points using groundwater are purposely designed to provide a secure water supply during drought. First, groundwater is less vulnerable in drought and provides a natural buffer during the first years of poor rains. Second, productive water points sited by test drilling in the greatest depths of saturated weathering will prevail as groundwater levels recede and other less well-sited water points fail. Third, the safe yield of a productive water point is based on a projection of the individual water point behaviour assuming an extended dry period of at least 240 days. Fourth, where manual pumps are used, they provide an inherent control against over-exploitation. Fifth, being community-based, a productive water point is often the single main point of abstraction in the area, often unaffected by abstraction from neighbouring

173

wells or boreholes and more readily managed than where many individual private wells are located on the same aquifer.

Together, these measures help to ensure that productive water point failure during drought is unlikely. Only where severe drought occurs at the end of an extended dry cycle (Figure 2.5) may additional control over the quantity of water abstracted become necessary. In this case, the water point must be managed so that the supply lasts until the next expected rains.

Experience shows that the management of productive water points must be based on consensus between users on the rules of the scheme – including the principle of payment for water supply, rights of access and special allocation arrangements during periods of drought. When it becomes necessary to control water use to prolong supply or to make a water point serve more people, a number of direct and indirect management actions can be taken by the water point committee. These include:

○ control leakage;
○ reuse water;
○ measure abstraction (e.g. fit water meters to pumps or counters on pump arms or do bucket counts to record daily abstraction) to reduce unaccountable losses and introduce quotas or differential tariffs for different water uses;
○ restrict pumping periods by locking the pumps or removing the handles for a part of each day;
○ use flow-limiting and water-saving devices, especially water-saving methods of irrigation, including irrigation late in the day;
○ reduce the irrigated area and grow crops that consume less water;
○ optimize conjunctive use of surface and ground water and use dual systems of higher and lower quality water where feasible;
○ install simple groundwater level monitors (e.g. floats and level indicators) to provide a graphic indication for the community of the water level remaining in the well or borehole;
○ use low discharge pumps (such as a windpump or solar pump) to lift water slowly but continuously to a tank when well water levels become low, and extend the sump to increase well storage, especially below the radials of collector wells;
○ reduce loss of rain as surface runoff and increase groundwater recharge in the surrounding catchment through land terracing, dead-level contours, small dams across ephemeral streams, recharge ponds, etc.

These management actions apply equally where over-development of the groundwater resource begins to cause interference between adjacent water points. Interpretation and provision of management information to the community members is vital. The users need to know how much water can be drawn each day. This can be facilitated by plotting water level and abstraction through time and should be a community initiative, undertaken by someone responsible who understands the concepts involved and who can make calculations concerning sustainable water use. Figure 7.8 is an example of a graph used to project sustainable water use at a productive water point in Zimbabwe during the severe drought of 1991/92. This sort of figure should be displayed publicly, ideally at the water point, and updated weekly as the drought progresses. The users must make a judgement as to when the next rains may come, but they can adjust water use accordingly. A strategic reserve of, say, 12 months drinking water for the whole community should be indicated, bringing all other water use back to a bare minimum for this period. In this way a safeguard is provided to ensure water is available if the drought lasts longer than anticipated.

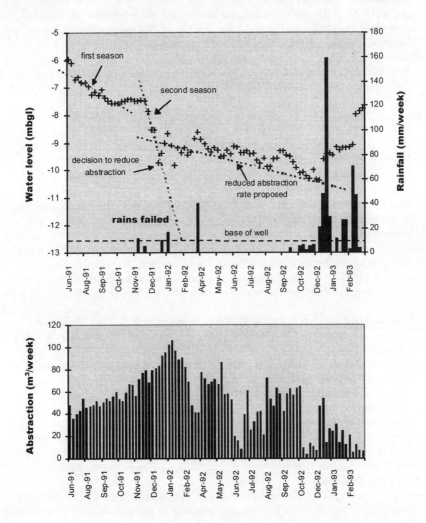

Figure 7.8 Drought management action plan, Romwe well 1991–92

References and further reading

Albertse, G. (1995). *Growing vegetables and grapes with the waggon wheel system.* Nietvoorbij Institute of Viticulture and Oenology, Stellenbosch, South Africa. 2 pp.

Al-Sakkaf, R.A., Zhou, Y. and Hall, M.J. (1999). A strategy for controlling groundwater depletion in the Sa'dah Plain, Yemen. *Water Resources Development* 15 (3) 349-365

Batchelor, C.H., Lovell, C.J. and Murata, M. (1996). Simple micro-irrigation techniques for improving irrigation efficiency on vegetable gardens. *Ag. Wat. Man.* 32, 37-48.

Bredehoeft, J. (1997) Safe yield and the water budget myth. Editorial, *Groundwater* 35(6) 929

Bredehoeft, J., Papadopulos, S.S. and Cooper, H.H. (1982). *Groundwater: The water-budget myth. Scientific basis of water resource management,* National Academy of Sciences Studies in Geophysics, 51-57.

Briggs, G.G. (1981). Theoretical and experimental relationships between soil adsorption, octanol-water partition coefficients, water solubilities, bioconcentration factors and the parachor. *Journal of Agriculture and Food Chemistry*, 29, 1050-1059.

Butterworth, J.A., Mugabe, F., Simmonds, L.P. and Hodnett, M.G. (1999) Hydrological processes and water resources management in a dryland environment II: Surface redistribution of rainfall within fields. *Hydrology and Earth System Sciences* 3(3), 333-343

Calow, R.C., Robins, N.S., Macdonald, A.M., Macdonald, D.M.J., Gibbs, B.R., Orpen, W.R.G., Mtembezeka, P., Andrews, A.J. and Appiah, S.O. (1997) Groundwater management in drought-prone areas of Africa. *Water Resources Development* 13(2) 241-261

Carter, R.C. and Howsam, P. (1994). Sustainable use of groundwater for small-scale irrigation. *Land Use Policy* 11(4), 275-285.

Chilton, P.J., Lawrence, A.R. and Stuart, M.E. (1995). The impact of tropical agriculture on groundwater quality. Chapter 10 in *Groundwater Quality* (Eds. Nash and McCall). Chapman and Hall, London. 113-122.

CIIFAD (1995). *Zimbabwe – Horticultural Crops Pest Management. Fact Sheets.* Cornell University International Institute of Food, Agriculture and Development, New York.

Custodio, E. (1989). Strict aquifer control rules versus unrestricted groundwater exploitation: comments on economic consequences. *Developments in Water Science* 39: 53-65

De Silva, C.S. and Rushton, K.R. (1996). Interpretation of the behaviour of agrowell systems in Sri Lanka using radial flow models. *Hydrological Sciences Journal* 41 (6) 825-835

De Silva, C.S. and Weatherhead, E.K. (1997). Optimising the dimensions of agrowells in hard rock aquifers in Sri Lanka. *Agricultural Water Management* 33, 117-126

De Silva, C.S., Fernando, N., Sakthivadivel, R. and Merrey, D. (1999). Managing groundwater in hard rock areas through agro-well design and development. *Water Resources Development* 15(3) 333-348.

Doorenbos, J. and Pruitt, W.O. (1977). Crop water requirements. *FAO Irrigation and Drainage Paper No. 24*, FAO, Rome. 114pp.

Elwell, H. and Maas, A. (1995). *Natural pest and disease control.* Natural Farming Network, Zimbabwe

Gore, K.P., Pendke, M.S., Gurunadha Rao, V.V.S. and Gupta, C.P. (1998). Groundwater modelling to quantify the effect of water harvesting structures in Wagarwadi watershed, Parbhani District, Maharashtra, India. *Hydrological Processes* 12, 1043-1052.

Hagmann, J. (1999) *Learning together for change: facilitating innovation in natural resource management through learning process approaches in rural livelihoods in Zimbabwe.* Margraf Verlag, Weikersheim, Germany. 330pp.

Huang, W.Y., Beach, E.D., Fernandez-Cornejo, J. and Uri, N.D. (1994). An assessment of the potential risks of groundwater and surface water contamination by agricultural chemicals used in vegetable production. *Science of the total environment,* 153, 151-167.

Kahnert, F. and Levine, G. (eds) (1993). *Groundwater irrigation and the rural poor: options for development in the Gangetic Basin.* World Bank, Washington, DC.

Kashililah, H. (1998). The WAMMA programme in Dodoma, Tanzania: a case study of community-managed pump and engine water supplies. In: *Water projects and livelihoods: poverty impact in a drought-prone environment.* Save the Children Fund Workshop Report No. 18, 74-76.

Kerr, J.M., Chandrakanth, M.G. and Deshpande, R.S. (1995) Economics of groundwater management in Karnataka. In: Kerr, J.M., Marothia, D.K., Singh, K., Ramasamy, C. and Bentley, W.R. (eds) *Natural resource economics: Concepts and applications to India.* Oxford and IBH. New Delhi.

Lovell, C.J., Murata, M., Batchelor, C.H. and Chilton, P.J. (1998b) Use of saline groundwater for community-based irrigation in dryland areas of Southern Africa. *Proc.10th ICID Afro-Asian Regional Conference in Bali, Indonesia, on Use of Saline and Brackish Water for Irrigation.* Eds. Ragab, R. and Pearce, G., pp. 106-115.

Murata, M. and Lovell, C.J. (1998). Garden irrigation: Alternative techniques and technology diffusion in dry regions. *Waterline*s 17 (2) 23-26.

Premanath, K.L.L. and Liyanapatabendi, T. (1994). Groundwater depletion due to agrowells. *Proc. 20th WEDC Conference: Affordable water supply and sanitation, Colombo*, 11-13.

Rao, P.S.C. and Alley, W.M. (1993). Pesticides. In: *Regional groundwater quality* (ed. W.M. Alley) Van Nostrand Reinhold, New York. 345-382.

Shah, T. (1989). Externality and equity implications of private exploitation of groundwater resources. *Developments in water science* 39, 459-482.

Shah, T. (1990). Sustainable development of groundwater resources: Lessons from Amrapur and Husseinabad Villages, India. *ODI/IIMI Irrigation Management Network Paper 90/3d*, ODI, Regent's College, London, 23pp.

Simmers, I., Villaroya, F. and Rebello, L.F. (eds) (1992). *Selected papers on aquifer overexploitation.* 23rd International Congress of the I.A.H. International Association of Hydrogeologists, vol.3, 391pp.

Singhal, B.B.S. (1997). Artificial recharge of groundwater in hard rocks with special reference to India. In: *Hard Rock Hydrosystems. Proc. Rabat Symposium.* IAHS Publ. No. 241, 11-12.

Sophocleous, M. (1997). Managing water resource systems: Why 'safe yield' is not sustainable. Editorial in *Groundwater* 35 (4), 561.

Theis, C.V. (1940). The source of water derived from wells: Essential factors controlling the response of an aquifer to development. *Civil Engineering* 10 (5) 277-280.

Villarroya, F. (1994). Regulatory issues mainly about aquifer over-exploitation within the scope of sustainable development. In: *Future groundwater resources at risk. Proc. Helsinki Conf. IAHS Publ. 222*, 389-402.

8 Anticipating the future of rural water supply

Productive water points and rural livelihood strategies

AN ESTIMATED 1.3 BILLION PEOPLE live in extreme poverty. Although increasing numbers of people live in towns and cities, the majority of people in developing countries still live in rural areas. Poverty remains disproportionately a rural phenomenon.

The 1997 UK Government White Paper on International Development commits the Department for International Development (DFID) to promoting 'sustainable rural livelihoods' and to promoting and improving the management of 'the natural and physical environment'. Both these objectives are expected to contribute to the goal of overall poverty eradication.

In support of this approach to poverty eradication, much effort has been put into refining a framework for analysis of livelihoods (see Figure 8.1). Livelihoods are defined as:

> 'comprising the capabilities, assets (including both material and social resources) and activities required for a means of living. A livelihood is sustainable when it can cope with and recover from stresses and shocks and maintain or enhance its capabilities and assets both now and in the future, while not undermining the natural resource base'.

The framework is holistic and dynamic, and recognizes the many complex interactions in rural livelihoods. It explicitly emphasizes the importance of institutions and organizations at all levels. It revolves around the five capital assets upon which individuals draw to build their livelihoods (Box 45). It helps promote consideration of the relationships between sector-wide approaches, decentralization and sustainable rural livelihoods, and the concepts of agricultural intensification, livelihood diversification and rural/urban linkages.

The framework provides a useful analytical tool both for looking at the impact of water projects and the scope for broader interventions to address poverty. We can see that productive water points alone will not modify the trend for rural people to migrate to already over-burdened cities in search of a better life. This will require a range of integrated development activities that seek to improve agriculture, home industries, transport, energy supply,

178

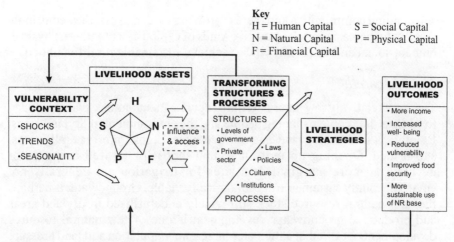

Figure 8.1 The Livelihoods Framework (adapted from Carney, 1998)

Box 45: Capital assets

Natural capital
The natural resources that are essential for livelihoods, e.g. land, water, wildlife, biodiversity, environmental resources.

Social capital
The social resources (networks, membership of groups, relationships of trust, access to wider institutions of society) upon which people draw in pursuit of their livelihoods.

Human capital
The skills, knowledge, ability to labour and good health important to pursue different livelihood strategies

Physical capital
The basic infrastructure (transport, shelter, water, energy and communications) and the production equipment and means which enable people to pursue their livelihoods.

Financial capital
The financial resources that are available to people (whether savings, supplies of credit or regular remittances or pensions) and which provide them with different livelihood options.

adapted from Scoones (1998)

---

health and education in rural areas. Nevertheless, productive water points have been found worthwhile even without these additional efforts because of their important contribution to improving water supply and local living conditions. Both groundwater and surface water-based schemes have a significant impact on the farming systems and livelihood strategies of the surrounding communities and there is growing evidence that they contribute directly to increasing each of the five kinds of capital asset (natural, physical, human, social and financial) vital to achieve sustainable rural livelihoods.

Natural capital

There are relatively few natural resource development options available today to address the problems facing people and the environment in dryland areas. Rainfall is unreliable, land and surface water stocks are generally over-subscribed. Opening up more marginal lands for rainfed cropping or developing more marginal dam sites for irrigation are generally not environmentally sustainable or economically viable. Groundwater is perhaps the only natural resource stock still widely underutilized in dryland areas and productive groundwater development is one of few natural resource development options that anticipates increasing population and land pressure and the need for agricultural intensification and livelihood diversification in the future. Although it is vital that increased groundwater use is carefully planned and closely monitored, but in many dryland areas a doubling of present use to allow production in conventional water supply programmes will have negligible impact on hydrology but enormous impact for rural poor in terms of improved health, nutrition, income and self-esteem. Approximately two-thirds of the world's dryland areas underlain by hard rock aquifers offer this potential for increased use to some degree.

Physical capital

An underlying principle in productive groundwater development is to make more effective use of existing infrastructure and to make new infrastructure more effective by siting and selecting well designs suited to the local groundwater conditions. An important benefit, principally for women, is saving time when fetching water and collecting vegetables. Time saving in rural water supply is not just a function of distance walked to a water point. It also depends on the water point functioning (i.e. being maintained), on the wait inevitable at existing water points of inappropriate design where water levels recover slowly after pumping, and on the queues common even at good water points where only a single handpump is fitted. Productive water points address each of these issues and provide relatively large amounts of water quickly through multiple pumps.

Human capital

The majority of garden members report that they have experienced positive improvements in their farming practices as a result of productive water points, mostly through the purchase of inputs to improve rainfed farm productivity. Importantly, this includes an increase in the renting of labour. Removal of labour as a constraint is also decreasing the seasonality of the costs and benefits at the margin of garden production, and growing and selling periods at the schemes are increasing. Non-garden members report that they have also experienced some improvement in their farming system. Both groups draw attention to the significant improvement in agricultural extension advice that has occurred, the productive water points providing a venue for increased social interaction. Other benefits cited include a strategic advantage in terms of the ability to buy seed and sow earlier in the season (using the winter income) thereby ensuring better yields and first crops at market, and the water charges now being made by some water point committees to pay for operation and maintenance, which are also beginning to be seen as a good source of income.

Social and financial capital

Respondents report that productive water points have given people the opportunity to share their experiences and skills, to work as a unit, and to pool ideas on different aspects so that they can succeed. Importantly, people are also expected, and are now able, to contribute money to these new ideas. Although relatively small in some cases, income from productive water points accrues to a wide spread of members. For those with limited access to cash or productive resources to start their own income-generating activities, obtaining a steady seasonal income from a productive water point lowers elements of risk and insecurity in the household budget and decision-making process. Some of the key constraints that households face in terms of getting involved with joint ventures in the community are eased, and the opportunities for entrepreneurial activity are increased.

This is reflected clearly in the amount of group activities and the size and success of savings clubs towards which scheme members feel they can now safely contribute some of their household income. A wide range of group or community-focused projects have started including projects that involve the purchase of equipment, the securing of credit facilities, and the accessing of 'matching-fund' type of rural development initiatives that can now be utilised. In terms of individual or household-based income-generating projects, the majority of members and a third of non-scheme members report that they have started a new project or improved an existing one at least in part as a direct result of a productive water point.

Surveys at standard (non-productive) domestic water points draw attention to the difference that a secure source of income from a productive water point can make in enhancing broader production systems. Although people at a standard domestic water point can design group-based activities as coherently as people at productive water points, neither the financial impact of the standard water point nor the implementation of new projects is prevalent. There remains a lack of collective action or sense of collective responsibility regarding ownership, management and development. Repairs still have to be undertaken by the state and support for new projects still has to come from a donor or local authority. The absence of an initial income-generating opportunity developed at the time the domestic water point was implemented appears to be critical. Without this initial income made available to many people, subsequent development of community-based activities is constrained and the productive capacity of the water point remains diffuse and underused. Any individual activities at the water point remain reactive to seasonal income needs and not pro-active in the generation of a financial surplus for the household or community to reserve or reinvest.

Integration into conventional rural water supply programmes

From a consideration of rural livelihoods we may expect that productive water points will play an important part in rural development in the future. In fact, many communities in dryland areas are already moving in this direction. Faced with diminishing holdings in crowded lands, rural people are digging more wells to start market gardens and other income-generating projects in a direct effort to intensify use of the land available to them and to diversify their livelihoods. It is important that conventional rural water supply programmes recognize this trend and adapt accordingly.

In some developing countries productive water points may fall between two 'institutional homes' because they combine rural water supply and agricultural development. However, the development priority will generally be to satisfy domestic water needs first and foremost, so it seems likely that productive water points will tend to be developed through the water sector rather than the agricultural sector in countries where these are separate.

Identifying what people actually want from their water points is critical and should underpin the design of future water supply programmes. If members of a community in a dryland area are asked an open question: *What do you want from a new water point?* they will invariably list a diverse range of productive activities that varies with their age and sex, in addition to a general call for a safe drinking water supply (Box 44). Increasingly, the list of productive activities will focus on income generation. In contrast, rural water supply has tended to focus only on improved access to domestic supply and improved sanitation. This needs to change, and the diverse range of small-scale production strategies that can be associated with a water point and which communities want, should be formally promoted.

Herein lies perhaps the greatest challenge. Today's rural water supply programmes have become largely prescriptive and follow a blueprint approach. Everything tends to be decided *a priori*, from the use the community will make of the water to the precise number and type of new water points that will be provided. Community choice and variable ground conditions are poorly catered for. Some official guidelines for rural water supply even stipulate the methods that can, and cannot, be used to site water points!

The demand responsive approach outlined in the previous chapters requires that consumers be engaged in the process of selecting, financing, implementing and managing water services that meet their needs and willingness to pay. The approach requires new ways of doing business, for all sector stakeholders – communities, NGOs, the private sector, government departments and donors. It requires a more flexible, adaptive approach that can evolve naturally, rather than the prescriptive approach currently taken in many programmes.

In meeting this challenge, it is important to focus on integrating productive water points and productive groundwater development into the existing programmes. Despite their prescriptive nature, there are still many similar and complementary steps between current water supply practice and productive groundwater development, and unique strengths in the respective approaches that will mutually reinforce one another if merged. The various agencies working on existing programmes have fairly well-defined roles and responsibilities, and a great deal of operational experience to draw on. For example, the Department of Water or its equivalent may have many years of experience in drilling boreholes, well sinking, pump installation, headwork construction, operation and maintenance and groundwater monitoring. The Department of Agricultural Extension may have trained field staff with experience of working with community farming groups. The Department of Health may have field staff trained in environmental health and nutrition education. The Department of Community Services may reach far into communities through village community workers whose focus is on primary health, hygiene education, income-generating activities, particularly for youth and women's groups, and social mobilization. The district councils may have a mandate as the key local development authority in their respective areas. Table 8.1 outlines the division of roles between these different actors consistent with the institutional framework of current water supply practice and current thinking on devolution of control down the various levels.

Government's role

Government has a critical role to play in creating an enabling environment for productive water point development and uptake of diverse production activities. This includes creating stable conditions that encourage other stakeholders to take a long-term view. At national level, government should:

O Develop policies and mechanisms that formally recognize and promote community-based irrigation and other productive activities, and which support integration of production and food security with rural water supply.

O Define or confirm a legal framework of land tenure, reinforcing traditional rights, where appropriate, which will serve as an incentive for the communities to invest in infrastructure for irrigation, livestock and other productive activities. This needs to include rights of access to groundwater under private and public land.

○ Develop policies and mechanisms to facilitate access to credit by community-based organizations.

○ Mobilize real-time information on markets and convey it to farmers' associations and facilitate the creation of farmer networks and marketing agents to disseminate and utilize such information.

○ Play a direct role in extension service training and provision of other technical support services, such as training on financial management, pest and disease control and marketing, and production of appropriate village-level training materials on these subjects.

○ Develop policies and mechanisms for state-wide implementation of groundwater monitoring and small watershed management systems.

○ Regulate potentially dangerous practices.

Table 8.1 Devolution of control in rural water supply programmes

ACTOR	ROLES
Communities	Negotiation of social contracts; pre-siting; formation of committees; provision of labour and local materials; ownership and management of water points and productive activities; maintenance protection of groundwater; implementation of catchment management plans.
Traditional leaders	Community leadership; social mobilization and consultation; local knowledge; conflict management; natural resource management.
District council (Executive and Councillors)	Overall administration, planning, co-ordination, management and monitoring of project components; development of district implementation plans; contracting out of certain activities; contract management. Policy making; social mobilization/consultation; liaison with traditional structures; approval of bye-laws; conflict management and arbitration.
Central government	In general: provision of policy framework and guidelines; training materials and resources; facilitatory support; technical support and supervision; monitoring and quality control.
Local government	Co-ordination of sector agencies; facilitatory support to district councils; progress monitoring and reporting.
Water development	Siting, drilling and pumping test tenders; pumping test tenders; contract supervision; pump installations; training communities in O&M; higher-level maintenance; monitoring of water points and groundwater resources.
Community services	Social mobilization/consultation; facilitation of social contracts, constitutions and bye-laws; facilitation of water point committee formation; organizational training; monitoring social processes.
Agricultural extension	Facilitating establishment of community gardens; education focusing on pest and disease control, crop rotations, marketing strategies, irrigation scheduling and water use efficiency; land use

	planning; village mapping; catchment management; arranging exposure visits and competitions; monitoring agricultural and environmental progress.
Health	Health and hygiene education; nutrition education and training; sanitation promotion; environmental health monitoring.
Natural resources	Training and support for local environmental planning and monitoring; links with district environmental action plans.
Forestry	Training and support for tree development in relation to environmental protection, catchment management and groundwater utilization.
Private sector	Responding to various tenders, such as pumping tests, upgrading existing water points, siting and constructing new water points; develop and manufacture new technologies; pump repair, headworks construction, pump lease arrangements, etc.
NGOs	Complementing government departments. Linking communities to sources of funding and information. Providing opportunities for testing new approaches and technologies. Providing guidance on lessons of best practice in social/participatory development processes. Appropriate technology development. Training resources for community capacity building.
Donors	Provision of financial and human resources; policy dialogue; facilitating independent evaluations; facilitating access to best practices from elsewhere.

At local authority level, interventions which may need to be piloted with communities, NGOs, government departments and private companies, but which are important for community-based development to be effective and sustainable, include:

o   orientation of local leaders and officials on the policy and guidelines for devolution of control for ownership and management of water points from local authority level to village government level.
o   registration of village government or community-based organizations responsible for this ownership and management.
o   formal agreement with the community on local and external contributions to the development, including available credit facilities.
o   registration of land tenure arrangements, including specific land allocations for schemes with clear compensation mechanisms and rights of access.
o   representation on pumping test appraisals as the customer checking on value for money.
o   formal approval of pump maintenance arrangements, including lease arrangements, between communities and private companies or commercialized government department.

- ○ gazetting the O&M partner mechanism and tariffs applicable.
- ○ arrange technical advisory services (TAS) to be provided e.g. information on safe yield and irrigable area, catchment protection, arbitration, finance mechanisms, pest and disease control, marketing strategies, pump technology, water distribution and irrigation methods.
- ○ gazetting the TAS partner mechanism and tariffs where applicable.
- ○ contracting out social development and extension tasks to local NGOs and private companies.
- ○ ongoing monitoring and evaluation to refine programme content and bye-laws.

Government does not have a positive role to play in building and managing productive water points. Experience shows that communities do much better when they build, own and operate the schemes themselves. From a centralized approach, where the government plays the role of supplier of water and sanitation, we need to move to a decentralized one, where the government acts as a promoter of a demand-responsive programme. In this approach, multi-sectoral teams can be set up. The main role of the district management team is to ensure that activities are implemented and to co-ordinate inter-ministerial collaboration. The multi-sectoral extension teams at area level are responsible for such things as participatory rural appraisal, community mobilization, civic education, water point committee formation and training, formulation of social contracts and constitutions, development and dissemination of training materials, monitoring and evaluation of project activities.

Donors' and NGOs' roles

Large donors move considerable sums of money which they are often in a hurry to disburse. The tendency has therefore been to support the large and visible state-supported irrigation systems. Donors need to focus on the smallholder systems (community-owned) which have greater impact on the poor and women, create more jobs and run at lower cost.

In some countries the application of appropriate water supply technology may be hindered by unduly restrictive legislative controls or overly-prescriptive implementation procedures. Donors should, therefore, encourage governments to review the legal and contractual frameworks, design standards and codes relevant to the water sector with a view to stimulating the use of appropriate technology. A new way of thinking may be needed, for example, by paying contractors per unit of water delivered rather than per hole in the ground.

Most institutions essential for effective running of smallholder irrigation and other productive activities centre around the community. Donors and NGOs should empower the communities rather than the bureaucrats, assisting the communities to acquire the water points from the state and to request technical support services from the state as a part of long-term community-based development plans. Donors and NGOs can respond to community efforts to strengthen the legal and institutional instruments of tenure and property rights that will allow greater say by communities over the land, water and other natural resources that they use, and yield greater investment and more efficient utilization.

There is need for donors and NGOs to encourage intensification of agriculture through support to microfinance schemes. Such schemes should be locally based and encompass both saving and borrowing. By supporting the production and marketing of high-value commodities, microfinance schemes, complete with marketing agents, could be a decisive factor for widespread and effective adoption of productive water points.

Donor and NGO support can strengthen the prospect and viability of productive water point uptake by providing mechanisms for information transfer. Pilot programmes may be needed to increase awareness of the available water supply technologies or where there is little experience of working in partnership with communities. Such pilot programmes could be established in a number of representative communities and would follow a learning-by-doing approach, providing training opportunities for both the private and public sectors, and acting as demonstration sites for local communities. Information obtained from these pilot programmes should be widely disseminated. This will help to avoid repetition and facilitate the application of improved systems. Research and development with the private sector of low-cost exploratory and horizontal drilling techniques may be fruitful. Finally, donors and NGOs can facilitate technology transfer, especially appropriate pump technologies, and can encourage local artisans to manufacture, market and carry spare parts for these pumps.

Legal requirements

Productive water points can contribute to the management of rural water supply, but the programme is still young and there is need to improve in a number of areas. Some of the special requirements that will help to integrate productive water points into conventional practice are summarized in Figure 8.2.

The success of pilot schemes in southern Africa is largely attributed to the independence of the farmers' groups in managing their own affairs - including water supply. This is not a new concept. For example, established irrigation boards in South Africa, which are legal entities, have been functioning along these principles for many decades. However, this is a new concept for rural water supply in many countries, where previously the state has played a major role in implementation and management.

Land ownership and the legal requirements to obtain permission to abstract groundwater for production are factors to consider. Water points on 'private' land may cause access problems for certain groups which may not be apparent

187

at the outset. Also, the potential problem of rent seeking by members at productive water points has been identified. During periods of water scarcity, water point committees are becoming increasingly concerned about the open access nature of the schemes, and it has been known for disincentive flat rate charges to be set to deter people from taking domestic water and for people who travel more than a certain distance to be turned away. As populations and livestock numbers increase, these pressures will only grow, and the definition of property rights to use the water will become increasingly important.

More research is needed to examine the extent of this problem and how it might develop over time. In some circumstances, the introduction of water rights, permits or pricing mechanisms may be beneficial in order to ensure that a more efficient and equitable system of capturing the benefits of the water is established.

Cost recovery

Preliminary willingness-to-pay analysis indicates that there is considerable potential for different groups to start new projects at productive water points by paying back loans. Future water supply programmes should investigate mechanisms whereby: revenue generated by a local water charging system can stay in and be managed by the local community; a proportion can still be used for recurrent operation and maintenance costs; payment by an individual or group entitles involvement in decision-making at the water point; and excess revenue generated by charging can best be utilized for production systems enhancement.

By way of example, surveys identify that over US$500 per year could potentially be collected from monthly payments at productive water points. This revenue could be used for routine operation and maintenance costs; a degree of cost recovery on the capital cost of the scheme; reinvestment in the garden or other group ideas; collateral for group loans; a credit pool from which micro-loans could be issued by the water point committee for further development of surrounding production systems and livelihood strategies. If designed and implemented carefully, the potential to capture and utilize these economic resources for the further benefit of the communities is both large and ultimately self-sustaining.

Integration into watershed development programmes

Collective action

It is becoming more accepted that the crucial issues of resource conservation and sustainable development can be addressed only if people enjoy a secure livelihood. Current development patterns and inequities increasingly force rural poor to subsist in marginal areas that have low productivity and that are ecologically fragile. Good and appropriate management of resources in these

188

Planning and logistics

Well designs are currently being chosen before drilling as the basis of contracts. The method has advantages, e.g. stockpiling of materials to reduce costs and delays; effective use of supervisory staff and equipment. The demand-responsive approach in which water points are chosen to suit local ground conditions poses important logistical challenges to ensure a cost-effective rolling programme.

Marketing

Lack of information is a major constraint to effective penetration of high-value markets by smallholder irrigators. State extension services rarely provide this information at present. Communities would value training on crop marketing strategies and rotations which take better account of market demands, and on diversification strategies that include other income-generating activities such as livestock feedlots, fish farming and dairying.

Institutional

Consolidate devolution of control for water and other natural resources to village level and continue to strengthen institutional capacities for developing and managing the resources at this level. Define or confirm a framework for village government that is clear on relations of power between local authority, traditional and political leaders, water point committee and other committees.
Provide training on financial management and record keeping to ensure proper accountability; business planning; marketing; contracting and negotiating skills; conflict resolution and leadership skills. Ensure that management of the resources by the community continues to be emphasized by encouraging communities to monitor their resources and determine safe annual harvests.

Technical

Standardize a pumping-test appraisal of yield that must be completed on all water points, including dug wells and existing boreholes. Perform the test in low groundwater conditions, ideally at the end of an extended dry cycle. Register all water points and consolidate database information to include at least: district and ward name; map co-ordinates; geology; date of pumping test; yield ($m^3 d^{-1}$); rest water level (m); total depth (m); first water strike (m); casing and screen details; principal water uses.
Train staff to update the database regularly and to use it to identify groundwater potential.
Expose drilling companies to the concepts of productive groundwater development, siting by test drilling, pumping test appraisal of existing and new water points, large diameter well construction and radial drilling. Stimulate adoption of well designs suited to local ground conditions by paying contractors per unit of water delivered rather than per hole in the ground, and avoid paying per metre depth drilled.

Special requirements to integrate productive water points into conventional rural water supply

Legal

Register community-based organizations as legal entities. Define or confirm a legal framework which transfers ownership and security of tenure for land, water and other natural resources to the local community and serves as incentive for the people to invest in productive water point infrastructure for irrigation, livestock, forestry and other productive activities.
Define or confirm a legal framework of water rights which will promote more efficient utilization and an equitable system of capturing the benefits of the water.

Economic and financial

Performance of groundwater irrigation systems can be improved by economic incentives to farmers, especially credit for the purchase of water distribution and irrigation technologies.
Encourage self-help development of new income-generating activities and maximize opportunities for involvement by promoting links with external agencies and matching-fund type programmes.
Subsidize construction for needy communities living in low potential areas where more expensive well designs will often be needed. Promote micro-finance schemes at district and village level. The success of revolving funds and loan repayments depends on proper management and the repayment capacity of the community in relation to system costs. Incentives to stimulate loan repayment, such as discounts for prompt payment, can be effective. Promote suitable banking facilities which provide sufficient interest to compensate for inflation.

Figure 8.2   Some special requirements to help integrate productive water points into conventional water supply practice

areas is crucial, if poverty and environmental degradation are to be avoided. Although no panacea, collective community action can play an important role in avoiding or reversing environmental degradation and in promoting sustainable development.

As earlier chapters have shown, the availability and distribution of groundwater resources in basement areas is such that setting up private productive water points for all members of a community is generally not viable. However, there are sufficient resources available for the implementation of reliable community schemes. What has yet to be determined is whether the experience gained by communities from implementing and managing their schemes (the enhanced social capital described earlier) can be channelled into other community-based projects such as community forestry or community feedlots aimed at reducing environmental degradation and promoting sustainable development. There is some reason to believe it can.

The underlying causes of environmental degradation are largely social, economic and institutional in origin. This means that the primary concern of community-based natural resource management should be the relationship between human welfare and a particular resource and its conservation for the use of future generations. The conservation of natural resources has to be linked to human welfare as the major motive force.

The basic social requirement to achieve this is that the operative unit, the producer/user community, should be small enough for households to participate meaningfully. The question of scale is critical for social cohesion. The smallest organization above the household – the village community – should be able to 'meet under the indaba tree' to decide management issues, as was customary in traditional open governance. If a community is too large or too dispersed for free discourse between members, it is preferable that it divides into smaller entities, each of which is then represented by a co-ordinating body. The economic requirement is that the producer/user community must benefit from their labour through the sale of their produce. This economic incentive provides the most important rationale for conserving the resource. The institutional requirements are: security of tenure for specific user groups; use of regulations evolved and enforced locally; and development and investment, or conservation and growth/expansion, of the resource in question. These three basic requirements (social, economic, institutional) of community-based natural resource management are each satisfied to some extent by the development of productive water points in dryland areas.

With increasing population and increasing reliance on groundwater for domestic use and production, the section of population at risk during low rainfall periods is growing. Collective community action will be increasingly important in the future, and the management of groundwater provides an important example of why this is so (Box 46). In areas where basement aquifers are reported to be failing (e.g. India) the common factor, apart from the already high population densities and significant increase in well numbers

190

Box 46: Collective management of groundwater

Collective management of basement aquifers in dryland areas will be increasingly important in the future. Unlike sedimentary or alluvial aquifers, basement aquifers are not an extensive homogenous resource but are localized, disjointed and spatially highly variable. There is rarely a regional water table to speak of. It must be recognized, therefore, that the areas of highest groundwater potential are limited. They will invariably lie under land already 'owned' by individual families and it will become increasingly important in water resource development to ensure equity of access to these resources. This can best be achieved by developing fewer high-yielding reliable community water points at these prime locations. Private wells may offer advantages in terms of individual management simplicity and add to general well being in times of plentiful rain, but they make groundwater management extremely difficult. Development of fewer well-sited community wells will have important advantages in this respect, and will represent a return to the collective management and collective wisdom.

in recent years, is the largely *ad hoc* way in which groundwater development has been allowed to occur on an individual or private basis. Inevitably, on the highly variable aquifers characteristic of basement, many of these water points are sited in less than optimum locations and are subsequently failing as groundwater levels fall through increased use and natural recession in low rainfall/recharge periods (Chapter 2). Although this failure is popularly attributed to "a regional decline in groundwater caused by overuse", the primary cause is the inappropriate development strategy adopted for the local groundwater conditions.

Taking India as an example of what may be in store for countries elsewhere in the future, it is clear that a strategy to develop fewer high-yielding communal water points at prime groundwater locations will have very important advantages in terms of groundwater management. Collective action of this form, rather than privatization by the better off, will also improve equity for those families without land or a private well, and will promote the local institutional structures needed to give the community one voice when negotiating with local authorities and development agencies for external support.

Catchment management

All the agenda-setting international meetings of recent years have led their recommendations by urging the adoption of sector strategies based on integrated water resources management (IWRM). In the case of rural water supply programmes, and productive water points in particular, that means integrating demand forecasts into plans for allocating water resources nationally; it means ensuring that proposals for productive activities are consistent with national strategies for water conservation and pollution prevention; it means institutional linkages to ensure compliance with water quality objectives and other environmental standards; and it means an increasing role for communities in catchment management.

Conclusions from a recent survey provide the best starting point when considering the role for communities in catchment management:

> Contrary to common viewpoints, the catchment or watershed is not always the most rational unit for all activities in rural development. Because neither catchments nor the groups who live among them are homogenous, the nature of their problems and the possible solutions are varied and complex. Prescriptive external solutions have little chance of fitting more than one-third of cases at best, and may be inappropriate or unacceptable to the majority of farmers. Nevertheless, working with common interest groups on contiguous areas of land, whose boundaries may be administrative, social or physical, enables agency staff to provide assistance more efficiently than where individual farms are scattered ...Thus it is not 'catchment management' as such that results in improvements in agriculture and livelihoods. Insistence on such a framework may run contrary to communities' needs and priorities ...Rather it is the integration of improved husbandry of land, of crops and of livestock with better interpersonal relations in the context of catchments that produces tangible benefits.

> *(Shaxson, 1999)*

Experience shows that catchment management is primarily about people, the users of the catchments themselves, far more than it is about hydrological 'catchments' or 'sub-catchments' that IWRM planners may demarcate on a map. It depends on community participation, which in turn depends on incentive. National IWRM programmes have generally failed to deliver, despite considerable interest and effort. The main reasons appear to be the top-down sectoral approach taken, a lack of community involvement in the process, and a lack of appropriate delivery mechanisms at ground level that generate the interest of local institutions.

There is a need, therefore, to identify approaches to implementing IWRM that engender active participation and reconcile current top-down (national) and bottom-up (community-based) approaches. Both are needed to achieve effective delivery in a structured programme beyond the scale of a few villages or isolated success stories. In developing countries, these approaches must also be appropriate to the socio-political setting at the same time as being affordable.

Formally linking IWRM to rural water supply, and to development of productive water points, is one strategy. Rural water supply is central to development in dry areas, already attracts donor funds, and brings together relevant ministries in an established institutional structure, which is important for scaling-up. Productive water points satisfy many of the criteria for an appropriate IWRM delivery mechanism. In addition to the obvious benefits derived, they also provide an initial step towards improved resource management at the wider scale. In many cases, they give rural communities their first experience of the institutions required for successful collective action. This, combined with the confidence gained from the initial project, can be a springboard to other community-based activities such as improved livestock management and forestry in the surrounding catchment.

It is important to note, however, that catchment stewardship does not happen spontaneously with the introduction of productive water points. Many issues influence the level of commitment that people in developing countries give to long-term environmental management initiatives. These include security of well-being, security of tenure, the increasing sub-division of land holdings, the management of common property resources, off-farm employment opportunities, the traditional right to return to rural homesteads, and the level of environmental education. These are issues that must also be considered by IWRM programmes.

Increased agricultural extension advice is reported at productive water points. Extension staff are capitalizing on the increased social interaction and this, combined with the local organizational structures developed, has increased general awareness of water resources management. However, surveys reveal that both the users and their extension staff have a poor understanding of hydrological processes and of the links between themselves, their activities in the catchment area, and the groundwater stored beneath their feet. Direct links need to be established between the natural resource base and the income being derived, ideally through environmental education.

A problem, however, is that catchment management programmes in dryland areas are generally being implemented on the basis of scant and often inappropriate information. Basic hydrological, agricultural and socio-economic data are often not available with which to define appropriate catchment management interventions or to develop the village-level extension materials needed. It is recommended that small (5-10 km^2) instrumented demonstration catchments representative of the major social and physical settings of interest should become an integral part of national IWRM programmes. This is a relatively small investment but will help to ensure that these national programmes do not end in disappointment.

Monitoring and evaluation

With increasing population pressure on natural resources will come increasing need to monitor and evaluate the sustainability and environmental trade-offs between different resource development options. But, ironically, some donor agencies are currently reducing monitoring and evaluation in a bid to reduce capital costs of development. An example in rural water supply is where pumping tests are no longer being commissioned at new water points. This is unwise. Resource management depends on information, and it is simply not possible to develop or manage a resource effectively if it is not quantified or monitored.

In fact sustainable cost-effective development of rural water supply depends on pumping test evaluation, projection of yields during drought, and routine water point monitoring. There is particular need to consolidate national database development in many countries. An important step towards this will be to standardize on a single pumping test appraisal that can be routinely applied to *all* water points: boreholes and wells, new and existing. This inventory should be performed during low groundwater conditions, ideally towards the end of an extended dry cycle and the information will form the basis of future drought management plans.

Routine monitoring of the groundwater resource can and should be a community initiative, performed on behalf of the state. This will allow the community to monitor the resource available to them and to foresee impending groundwater drought (Chapter 7). Establishing monitoring networks costs money, and the development of low-cost methods which communities can use to record abstraction and monitor groundwater levels will be important.

Whether a newly constructed water point or an up-graded water point, there is no better 'pumping test' than the long-term monitoring records kept by the communities themselves. As the available information increases, increasing confidence can be placed on the water supply available to them and the appropriate management options.

Concluding remarks

The primary purpose of this book has been to set out the principles, procedures and practices that should guide decisions on the choice, design and management of productive water points in dryland areas overlying hard rock aquifers. Because the effectiveness and sustainability of rural water supply depends not only on technology choice, but also, critically, on user involvement, enabling legislation, the right partnership approaches, innovative community-based financing, and the promotion of appropriate technical support services, the guidance has included discussion of social, financial, legal and institutional aspects as well as technical, agricultural and environmental concerns.

However, rural water supply is in a state of transition in many dryland areas, trying to come to terms with rapid social, political and economic changes, population growth and growing demands on a scarce resource. Where negotiated approaches involving the users themselves are now recognized to ensure more equitable, efficient and acceptable outcomes, the package of interventions and ideas presented in this book may be useful. Elsewhere, perhaps where community participation remains minimal and where rural water supply is still state-implemented and maintained, only the technical details of available water supply options may be of interest. There are important differences between countries, but I hope that you are able to draw something from this text that you find useful and appropriate to your own setting.

References and further reading

Adams, M., Sibanda, S. and Turner, S. (1999). Land tenure reform and rural livelihoods in Southern Africa. *Natural Resource Perspectives No. 39*, ODI, London.

Blackmore, D.J. (1995). Murray-Darling Basin Commission: a case study in integrated catchment management. *Water Science and Technology* 32, 15-25.

Carney, D. (ed) (1998). *Sustainable rural livelihoods: what contribution can we make?* DFID, London. 213 pp.

Ellis, F. (1999). Rural livelihood diversity in developing countries: evidence and policy implications. *Natural Resource Perspectives No. 40*, ODI, London.

Farrington, J. and Boyd. C. (1997). Scaling up the participatory management of common pool resources. *Development Policy Review 15*, 371-391.

Farrington, J. and Lobo, C. (1997) Scaling up participatory catchment development in India: lessons from the Indo-German Watershed Development Programme. *ODI Natural Resource Perspectives No. 17*, 6pp.

Kahnert, F. and Levine, G. (eds.) (1993). *Groundwater irrigation and the rural poor: Options for development in the Gangetic Basin.* The World Bank, Washington.

Juma, C. and Ojwang, J.B. (1996). *In land we trust – environment, private property and constitutional change.* Initiatives Publ., Nairobi and Zed Books, London.

Ostrom, E. (1995). Designing complexity to govern complexity. In: Hanna, S. and Munasinghe, M. (Eds) *Property rights and the environment: Social and ecological issues.* Washington DC. The Beijer International Institute of Ecological Economics and the World Bank. 33-45.

Rhoades, R.E. (1998). Participatory watershed research and management: where the shadow falls. *Gatekeeper Series No. SA81.* International Institute for Environmental Development.

Scoones, I. (1998). Sustainable rural livelihoods: a framework for analysis. *Working Paper No. 72.* Institute of Development Studies, Brighton.

Shaxson, F. (1999). Foreword to: *Fertile Ground – the impacts of participatory watershed management.* (eds. Hinchcliffe, F., Thompson, J., Pretty, J., Guijt, I. and Shah, P.) Intermediate Technology Publications, London.

Turton, C. and Farrington, J. (1998) Enhancing rural livelihoods through participatory watershed development in India. *Natural Resource Perspectives No. 34*, ODI, London.

van Zyl, F.C. (1995). Integrated catchment management: is it wishful thinking or can it succeed? *Water Science and Technology 32*, 27-35.

Warner, M. and Jones, P. (1998). Assessing the need to manage conflict in community-based natural resource projects. *Natural Resource Perspectives No. 35*, ODI, London.

Whiteside, M. (1998). Encouraging sustainable smallholder agriculture in Southern Africa in the context of agricultural services reform. *Natural Resource Perspectives No. 36*, ODI, London.

Wily, L. (1999). Moving forward in African community forestry: Trading power, not use rights. *Society and Natural Resources 12*, 49-61.

Appendix 1: Cost/benefit analysis of 13 productive water point options

Productive water point option		1 Upgrade under-used small dam		2 Upgrade under-used borehole		3 Convert to screened regolith borehole	
		Multiple manual pumps	Motor pump	Multiple manual pumps	Motor pump	Multiple manual pumps	Motor pump
Capital Costs (1994 Z$)	% imported						
Site ID by records	0	0	0	500	500	500	500
Site ID by exploratory drilling	100	0	0	0	0	0	0
Site ID by geophysics	100	0	0	0	0	0	0
Pumping tests/modelling yield	0	0	0	4800	4800	2400	2400
Vertical drilling	100	0	0	0	0	14 000	14 000
Dig well or dam construction	50	130 000	130 000	0	0	0	0
Horizontal drilling (radials)	100	0	0	0	0	0	0
Lining and headworks	50	0	0	1000	1000	2500	2500
Sub-Total (1994 Z$)		**130 000**	**130 000**	**6300**	**6300**	**19 400**	**19 400**
Real devaluation inflator:	1.6089	CSO & RBZ					
CPI inflator:	2.7709	Idem.					
Profit inflator PWP only	1.1	Assumed					
Exchange rate, Dec 1998, US$:Z$	37.36	Idem.					
Capital Costs (Dec 1998 US$)	Devaluation						
taking account of devaluation	multiplier						
Site ID by records	1.0000	0	0	41	41	41	41
Site ID by exploratory drilling	1.6089	0	0	0	0	0	0
Site ID by geophysics	1.6089	0	0	0	0	0	0
Pumping tests/modelling yield	1.0000	0	0	392	392	196	196
Vertical drilling	1.6089	0	0	0	0	1838	1838
Dug well or dam construction	1.3045	13 835	13 835	0	0	0	0
Horizontal drilling (radials)	1.6089	0	0	0	0	0	0
Lining and headworks	1.3045	0	0	106	106	266	266
Sub-total (water point)		**13 835**	**13 835**	**539**	**539**	**2340**	**2340**
Average yield of water (m³ day⁻¹)		100.0	100.0	44.0	44.0	36.4	36.4
Pump type		treadle	centrifug	hydro	mono	hydro	mono
Single pump capacity (m³ day⁻¹)		36.0	100.0	11.0	44.0	11.0	36.4
Single pump cost (US$)	1.0000	150	900	800	2300	800	2300
No. pumps to reach capacity		3	1	4	1	4	1
Sub-total (pumps)		**450**	**900**	**3200**	**2300**	**3200**	**2300**
Fenced area (ha)	0.032 x yield	3.20	3.20	1.41	1.41	1.16	1.16
Value of land	$US 100/ha	320	320	141	141	116	116
Cost of fence	$US 820/ha	2624	2624	1155	1155	955	955
Sub-total (garden)		**2944**	**2944**	**1295**	**1295**	**1072**	**1072**
Sub-total (social development)	$US 1000/site	**1000**	**1000**	**1000**	**1000**	**1000**	**1000**
Total Capital Investment		**18 229**	**18 679**	**6034**	**5134**	**7612**	**6712**
Recurrent Costs (Dec 1998 US$)							
Depreciation: waterpoint & garden	20 yrs; 5% p.a.	839	839	92	92	171	171
Depreciation: pumps	10 yrs; 10% p.a	45	90	320	230	320	230
Pump repairs	5% pump cost	23	45	160	115	160	115
Fuel	US$0.05 m⁻³	0	1200	0	528	0	437
Labour	US$0.23 / 8hr day	4141	2483	1822	1093	1507	904
Total		**5047**	**4657**	**2394**	**2057**	**2158**	**1856**
Garden benefits (Dec 1998 US$)	input GM/ha						
Average Gross Margin (US$/year)	2215	7088	7088	3119	3119	2580	2580

	Year	Upgrade under-used small dam		Upgrade under-used borehole		Convert to screened regolith borehole	
		Multiple manual pumps	Motor pump	Multiple manual pumps	Motor pump	Multiple manual pumps	Motor pump
Cashflow with labour costed	Year	-18 229	-18 679	-6034	-5134	-7612	-6712
(Dec 1998 US$/year)	1	2041	2431	725	1061	422	724
recurrent costs + 13% inflation	2	2306	2747	819	1199	477	818
garden benefits + 13% inflation	3	2606	3104	926	1355	539	924
	4	2944	3507	1046	1531	609	1044
	5	3327	3963	1182	1731	688	1180
	6	3760	4478	1336	1955	778	1333
	7	4248	5061	1509	2210	879	1507
	8	4801	5719	1706	2497	993	1703
	9	5425	6462	1927	2822	1122	1924
	10	6130	7302	2178	3188	1268	2174
Replace pumps	11	6927	8251	2461	3603	1433	2457
	12	7828	9324	2781	4071	1619	2776
	13	8845	10 536	3143	4600	1830	3137
	14	9995	11 906	3551	5199	2068	3545
	15	11 294	13 453	4013	5874	2336	4006
	16	12 763	15 202	4534	6638	2640	4526
	17	14 422	17 179	5124	7501	2983	5115
	18	16 297	19 412	5790	8476	3371	5780
	19	18 415	21 935	6543	9578	3809	6531
	20	20 809	24 787	7393	10 823	4305	7380
IRR years 1-20 (% p.a.)		**22**	**24**	**23**	**33**	**13**	**21**
Net Present Value (13%)		**17 888**	**24 342**	**6798**	**13 651**	**-141**	**6097**
Cashflow with labour costed	Year	-18229	-18679	-1315	-415	-2893	-1993
relative to standard domestic	1	2041	2431	1010	1346	707	1009
borehole (Dec 1998 US$/year)	2	2306	2747	1104	1484	762	1103
recurrent costs + 13% inflation	3	2606	3104	1211	1640	824	1209
garden benefits + 13% inflation	4	2944	3507	1331	1816	894	1329
	5	3327	3963	1467	2015	973	1465
	6	3760	4478	1621	2240	1063	1618
	7	4248	5061	1794	2495	1164	1792
	8	4801	5719	1991	2782	1278	1988
	9	5425	6462	2212	3106	1407	2209
	10	6130	7302	2463	3473	1553	2459
Replace pumps	11	6927	8251	2746	3888	1718	2742
	12	7828	9324	3066	4356	1904	3061
	13	8845	10 536	3427	4885	2115	3422
	14	9995	11 906	3836	5483	2353	3830
	15	11 294	13 453	4298	6159	2621	4290
	16	12 763	15 202	4819	6923	2925	4811
	17	14 422	17 179	5409	7786	3268	5400
	18	16 297	19 412	6075	8761	3656	6065
	19	18 415	21 935	6828	9863	4094	6816
	20	20 809	24 787	7678	11 108	4590	7665
IRR years 1-20 (% p.a.)		**22**	**24**	**87**	**334**	**33**	**61**
Net Present Value (13%)		**17 888**	**24 342**	**13 518**	**20 371**	**6580**	**12 817**

Over 20 year life assuming 13% inflation:

	Multiple manual pumps	Motor pump	Multiple manual pumps	Motor pump	Multiple manual pumps	Motor pump
Total benefits US$	573 741	573 741	252 446	252 446	208 842	208 842
Cost to community without interest but labour costed	426 789	395 662	199 793	171 667	182 283	156 972
Cost to community without interest or labour costs	91 604	194 654	52 312	83 224	60 276	83 806
Benefit/cost ratio (with labour costs)	1.3	1.5	1.3	1.5	1.1	1.3
Benefit/cost ratio (without labour costs)	6.3	2.9	4.8	3.0	3.5	2.5
Total cost US$ / m³ of water (with labour costs)	0.89	0.82	0.95	0.81	1.04	0.90
Total cost US$ / m³ of water (without labour costs)	0.19	0.41	0.25	0.39	0.34	0.48
Total cost US$ / ha of irrigation (with labour costs)	6669	6182	7095	6096	7825	6738
Total cost US$ / ha of irrigation (without labour costs)	1431	3041	1858	2955	2587	3597
Per capita capital cost US$ (80 families at 9.3 people)	24.50	25.11	8.11	6.90	10.23	9.02
Per capita recurrent cost US$/yr (with labour costs)	6.78	6.26	3.22	2.77	2.90	2.50
Per capita recurrent cost US$/yr (without labour costs)	1.22	2.92	0.77	1.30	0.87	1.28

Productive water point option

		4 Convert to large-diameter well		**5** Convert to collector well		**6** Site and construct new small dam	
		Multiple manual pumps	Motor pump	Multiple manual pumps	Motor pump	Multiple manual pumps	Motor pump
Capital Costs (1994 Z$)	% imported						
Site ID by records	0	500	500	500	500	0	0
Site ID by exploratory drilling	100	0	0	0	0	0	0
Site ID by geophysics	100	0	0	0	0	0	0
Pumping tests/modelling yield	0	3600	3600	4800	4800	0	0
Vertical drilling	100	0	0	0	0	0	0
Dig well or dam construction	50	20 100	20 100	20 100	20 100	1 700 000	1 700 000
Horizontal drilling (radials)	100	0	0	15 500	15 500	0	0
Lining and headworks	50	11 450	11 450	19 900	19 900	0	0
Sub-Total (1994 Z$)		**35 650**	**35 650**	**60 800**	**60 800**	**1 700 000**	**1 700 000**
Real devaluation inflator:	1.6089	CSO & RBZ					
CPI inflator:	2.7709	Idem.					
Profit inflator PWP only	1.1	Assumed					
Exchange rate, Dec 1998, US$:Z$	37.36	Idem.					
Capital Costs (Dec 1998 US$) taking account of devaluation	Devaluation multiplier						
Site ID by records	1.0000	41	41	41	41	0	0
Site ID by exploratory drilling	1.6089	0	0	0	0	0	0
Site ID by geophysics	1.6089	0	0	0	0	0	0
Pumping tests/modelling yield	1.0000	294	294	392	392	0	0
Vertical drilling	1.6089	0	0	0	0	0	0
Dug well or dam construction	1.3045	2139	2139	2139	2139	180 922	180 922
Horizontal drilling (radials)	1.6089	0	0	2035	2035	0	0
Lining and headworks	1.3045	1219	1219	2118	2118	0	0
Sub-total (water point)		**3692**	**3692**	**6724**	**6724**	**180 922**	**180 922**
Average yield of water (m³ day⁻¹)		25.1	25.1	29.5	29.5	100.0	100.0
Pump type		bush	mono	bush	mono	treadle	centrifug
Single pump capacity (m³ day⁻¹)		18.0	25.1	18.0	29.5	36.0	100.0
Single pump cost (US$)	1.0000	245	2300	245	2300	150	900
No. pumps to reach capacity		2	1	2	1	3	1
Sub-total (pumps)		**490**	**2300**	**490**	**2300**	**450**	**900**
Fenced area (ha)	0.032 x yield	0.80	0.80	0.94	0.94	3.20	3.20
Value of land	$US 100/ha	80	80	94	94	320	320
Cost of fence	$US 820/ha	659	659	774	774	2624	2624
Sub-total (garden)		**739**	**739**	**868**	**868**	**2944**	**2944**
Sub-total (social development)	$US 1000/site	**1000**	**1000**	**1000**	**1000**	**1000**	**1000**
Total Capital Investment		**5921**	**7731**	**9082**	**10 892**	**185 316**	**185 766**
Recurrent Costs (Dec 1998 US$)							
Depreciation: waterpoint & garden	20 yrs; 5% p.a.	222	222	380	380	9193	9193
Depreciation: pumps	10 yrs; 10% p.a	49	230	49	230	45	90
Pump repairs	5% pump cost	25	115	25	115	23	45
Fuel	US$0.05 m⁻³	0	301	0	354	0	1200
Labour	US$0.23 / 8hr day	1039	623	1222	733	4141	2483
Total		**1334**	**1491**	**1675**	**1811**	**13 402**	**13 011**
Garden benefits (Dec 1998 US$)	input GM/ha						
Average Gross Margin (US$/year)	2215	1779	1779	2091	2091	7088	7088

	Year	Convert to large-diameter well		Convert to collector well		Site and construct new small dam	
		Multiple manual pumps	Motor pump	Multiple manual pumps	Motor pump	Multiple manual pumps	Motor pump
Cashflow with labour costed	Year	−5921	−7731	−9082	−10 892	−185 316	−185 766
(Dec 1998 US$/year)	1	445	288	416	280	−6314	−5924
recurrent costs + 13% inflation	2	502	325	470	316	−7134	−6694
garden benefits 13% inflation	3	568	368	532	357	−8062	−7564
	4	642	416	601	404	−9110	−8547
	5	725	470	679	456	−10 294	−9658
	6	819	531	767	515	−11 633	−10 914
	7	926	600	887	582	−13 145	−12 333
	8	1046	678	979	658	−14 854	−13 936
	9	1182	766	1107	744	−16 785	−15 748
	10	1336	865	1250	840	−18 967	−17 795
Replace pumps	11	1509	978	1413	950	−21 432	−20 108
	12	1706	1105	1597	1073	−24 219	−22 722
	13	1927	1248	1804	1213	−27 367	−25 676
	14	2178	1411	2039	1370	−30 925	−29 014
	15	2461	1594	2304	1548	−33 945	−32 786
	16	2781	1801	2603	1750	−39 488	−37 048
	17	3143	2036	2942	1977	−44 621	−41 864
	18	3551	2300	3324	2234	−50 422	−47 306
	19	4013	2599	3756	2525	−56 977	−53 456
	20	4534	2937	4245	2853	−64 384	−60 406
IRR years 1-20 (% p.a.)		**16**	**9**	**11**	**6**	**negative**	**negative**
Net Present Value (13%)		**1949**	**−2633**	**−1715**	**−5941**	**−297 063**	**−290 609**
Cashflow with labour costed	Year	−1202	−3012	−4364	−6174	−185 316	−185 766
relative to standard domestic	1	730	573	701	565	−6314	−5924
borehole (Dec 1998 US$/year)	2	787	610	755	601	−7134	−6694
recurrent costs + 13% inflation	3	853	653	816	642	−8062	−7564
garden benefits + 13% inflation	4	927	701	886	689	−9110	−8547
	5	1010	755	964	741	−10 294	−9658
	6	1104	816	1052	800	−11 633	−10 914
	7	1211	885	1152	867	−13 145	−12 333
	8	1331	963	1264	943	−14 854	−13 936
	9	1467	1051	1392	1029	−16 785	−15 748
	10	1621	1150	1535	1125	−18 967	−17 795
Replace pumps	11	1794	1263	1698	1235	−21 432	−20 108
	12	1991	1390	1882	1358	−24 219	−22 722
	13	2212	1533	2089	1498	−27 367	−25 676
	14	2463	1696	2324	1655	−30 925	−29 014
	15	2746	1879	2589	1833	−34 945	−32 786
	16	3066	2086	2888	2035	−39 488	−37 048
	17	3428	2320	3227	2262	−44 621	−41 864
	18	3836	2585	3609	2519	−50 422	−47 306
	19	4298	2884	4041	2810	−56 977	−53 456
	20	4819	3222	4530	3138	−64 384	−60 406
IRR years 1–20 (%p.a.)		**69**	**26**	**24**	**14**	**negative**	**negative**
Net Present Value (13%)		**8669**	**4087**	**5006**	**779**	**−297 063**	**−290 609**

Over 20 year life assuming 13% inflation:

Total benefits US$	144 009	144 009	169 254	169 254	573 741	573 741
Cost to community without interest but labour costed	113 936	128 426	144 641	157 501	1 270 132	1 239 005
Cost to community without interest or labour costs	29 805	77 973	45 761	98 204	934 948	1 037 998
Benefit/cost ratio (with labour costs)	1.3	1.1	1.2	1.1	0.5	0.5
Benefit/cost ratio (without labour costs)	4.8	1.8	3.7	1.7	0.6	0.6
Total cost US$/m^3 of water (with labour costs)	0.95	1.07	1.02	1.11	2.65	2.58
Total costs US$/m^3 of water (without labour costs)	0.25	0.65	0.32	0.69	1.95	2.16
Total cost US$ / ha of irrigation (with labour costs)	7093	7995	7661	8342	19 846	19 359
Total cost US$ / ha of irrigation (without labour costs)	1855	4854	2424	5201	14 609	16 219
Per capita capital cost US$ (80 families at 9.3 people)	7.96	10.39	12.21	14.64	249.08	249.69
Per capita recurrent cost US$/yr (with labour costs)	1.79	2.00	2.25	2.43	18.01	17.49
Per capita recurrent cost US$/yr (without labour costs)	0.40	1.17	0.61	1.45	12.45	14.15

Productive water point option		7 New screened regolith borehole		8 New large-diameter well		9 New collector well	
		Multiple manual pumps	Motor pump	Multiple manual pumps	Motor pump	Multiple manual pumps	Motor pump
Capital costs (1994 Z$)	% imported						
Site ID by records	0	0	0	0	0	0	0
Site ID by exploratory drilling	100	12 000	12 000	12 000	12 000	12 000	12 000
Site ID by geophysics	100	0	0	0	0	0	0
Pumping tests/modelling yield	0	2400	2400	3600	3600	4800	4800
Vertical drilling	100	14 000	14 000	0	0	0	0
Dig well or dam construction	50	0	0	20 100	20 100	20 100	20 100
Horizontal drilling (radials)	100	0	0	0	0	15 500	15 500
Lining and headworks	50	2500	2500	11 450	11 450	19 900	19 900
Sub-total (1994 Z$)		**30 900**	**30 900**	**47 150**	**47 150**	**72 300**	**72 300**
Real devaluation inflator:	1.6089	CSO & RBZ					
CPI inflator:	2.7709	Idem.					
Profit inflator PWP only	1.1	Assumed					
Exchange rate, Dec 1998, US$:Z$	37.36	Idem.					
Capital costs (Dec 1998 US$) taking account of devaluation	Devaluation multiplier						
Site ID by records	1.0000	0	0	0	0	0	0
Site ID by exploratory drilling	1.6089	1575	1575	1575	1575	1575	1575
Site ID by geophysics	1.6089	0	0	0	0	0	0
Pumping tests/modelling yield	1.0000	196	196	294	294	392	392
Vertical drilling	1.6089	1838	1838	0	0	0	0
Dug well or dam construction	1.3045	0	0	2139	2139	2139	2139
Horizontal drilling (radials)	1.6089	0	0	0	0	2035	2035
Lining and headworks	1.3045	266	266	1219	1219	2118	2118
Sub-total (water point)		**3875**	**3875**	**5227**	**5227**	**8258**	**8258**
Average yield of water (m³/day)		36.4	36.4	25.1	25.1	29.5	29.5
Pump type		hydro	mono	bush	mono	bush	mono
Single pump capacity (m³/day)		11.0	36.4	18.0	25.1	18.0	29.5
Single pump cost (US$)	1.0000	800	2300	245	2300	245	2300
No. of pumps to reach capacity		4	1	2	1	2	1
Sub-total (pumps)		**3200**	**2300**	**490**	**2300**	**490**	**2300**
Fenced area (ha)	0.032 × yield	1.16	1.16	0.80	0.80	0.94	0.94
Value of land	US$ 100/ha	116	116	80	80	94	94
Cost of fence	US$ 820ha	955	955	659	659	774	774
Sub-total (garden)		**1072**	**1072**	**739**	**739**	**868**	**868**
Sub-total(social development)	US$ 1000/site	**1000**	**1000**	**1000**	**1000**	**1000**	**1000**
Total Capital Investment		**9146**	**8246**	**7456**	**9266**	**10 617**	**12 427**
Recurrent costs (Dec 1998 US$)							
Depreciation: waterpoint & garden	20 yrs; 5% p.a.	247	247	298	298	456	456
Depreciation: pumps	10 yrs; 10% p.a.	320	230	49	230	49	230
Pump repairs	5% pump cost	160	115	25	115	25	115
Fuel	US$0.05m⁻³	0	437	0	301	0	354
Labour	US$0.23 / 8hr day	1507	904	1039	623	1222	733
Total		**2235**	**1933**	**1411**	**1568**	**1751**	**1888**
Garden benefits (Dec 1998 US$)	input GM/ha						
Average Gross Margin (US$/year)	2215	2580	2580	1779	1779	2091	2091

		New screened regolith borehole		New large-diameter well		New collector well	
		Multiple manual pumps	Motor pump	Multiple manual pumps	Motor pump	Multiple manual pumps	Motor pump
Cashflow with labour costed	Year	−9146	−8246	−7456	−9266	−10 617	−12 427
(Dec 1998 US$/year)	1	345	647	368	211	340	203
recurrent costs + 13% inflation	2	390	731	416	239	384	229
garden benefits + 13% inflation	3	441	826	470	270	434	259
	4	498	934	531	305	490	293
	5	563	1055	600	345	554	331
	6	636	1192	678	389	626	374
	7	719	1347	766	440	707	423
	8	813	1522	866	497	799	478
	9	918	1720	978	562	903	540
	10	1038	1944	1105	635	1020	610
Replace pumps	11	1173	2196	1249	717	1153	689
	12	1325	2482	1411	811	1302	779
	13	1497	2804	1595	916	1472	880
	14	1692	3169	1802	1035	1663	994
	15	1912	3581	2036	1169	1879	1124
	16	2160	4046	2301	1322	2124	1270
	17	2441	4572	2600	1493	2400	1435
	18	2759	5167	2938	1687	2712	1621
	19	3117	5839	3329	1907	3064	1832
	20	3522	6598	3752	2155	3462	2070
IRR years 1-20 (% p.a.)		**9**	**17**	**12**	**5**	**7**	**2**
Net Present Value (13%)		**−3033**	**3205**	**−943**	**−5526**	**−4607**	**−8833**
Cashflow with labour costed	Year	−4428	−3528	−2737	−4547	−5898	−7708
relative to standard domestic	1	630	932	653	496	624	488
borehole (Dec 1998 US$/year)	2	675	1016	701	524	669	514
recurrent costs + 13% inflation	3	726	1111	755	555	719	544
gardens benefits +13% inflation	4	783	1218	816	590	775	578
	5	848	1340	885	629	839	616
	6	921	1477	963	674	911	659
	7	1004	1632	1051	725	992	708
	8	1098	1807	1151	782	1084	763
	9	1203	2005	1263	847	1188	825
	10	1323	2229	1390	920	1305	895
Replace pumps	11	1457	2481	1534	1002	1438	974
	12	1610	2767	1696	1095	1587	1064
	13	1782	3089	1880	1201	1757	1165
	14	1977	3454	2087	1320	1948	1279
	15	2197	3866	2321	1454	2164	1409
	16	2445	4331	2586	1606	2409	1555
	17	2726	4857	2885	1778	2685	1720
	18	3043	5452	3223	1972	2997	1906
	19	3402	6124	3605	2192	3349	2117
	20	3807	6883	4037	2440	3747	2355
IRR years 1-20 (% p.a.)		**21**	**36**	**32**	**16**	**17**	**9**
Net Present Value (13%)		**3688**	**9925**	**5777**	**1195**	**2113**	**−2113**

Over 20 year life assuming 13% inflation:

	New screened regolith borehole		New large-diameter well		New collector well	
	Multiple manual pumps	Motor pump	Multiple manual pumps	Motor pump	Multiple manual pumps	Motor pump
Total benefits US$	208 842	208 842	144 009	144 009	169 254	169 254
Cost to community without interest but labour costed	190 028	164 717	121 681	136 171	152 385	165 245
Costs to community without interest or labour costs	68 020	91 550	37 550	85 718	53 506	105 948
Benefit/cost ratio (with labour costs)	1.1	1.3	1.2	1.1	1.1	1.0
Benefit/cost ratio (without labour costs)	3.1	2.3	3.8	1.7	3.2	1.6
Total cost US$/m^3 of water (with labour costs)	1.09	0.94	1.01	1.13	1.08	1.17
Total cost US$/m^3 of water (without labour costs)	0.39	0.52	0.31	0.71	0.38	0.75
Total cost US$/ha of irrigation (with labour costs)	8157	7071	7575	8477	8071	8752
Total cost US$/ha of irrigation (without labour costs)	2920	3930	2338	5336	2834	5612
Per capita capital cost US$ (80 families at 9.3 people)	12.29	11.08	10.02	12.45	14.27	16.70
Per capita recurrent cost US$/yr (with labour costs)	3.00	2.60	1.90	2.11	2.35	2.54
Per capita recurrent costs US$/yr (without labour costs)	0.98	1.38	0.50	1.27	0.71	1.55

Productive water point option

		10 Hydro-fracture existing borehole	11 New conv. deep borehole	12 Deep well av. depth 17.6 m	13 Family well av. depth 10.8 m	Standard domestic borehole to 55 m
Capital costs (1994 Z$)	% imported					
Site ID by records	0	0	0	0	0	0
Site ID by exploratory drilling	100	0	0	0	0	0
Site ID by geophysics	100	0	1000	0	0	1000
Pumping tests/modelling yield	0	0	0	0	0	0
Vertical drilling	100	6516	30 000	0	0	30 000
Dig well or dam construction	50	0	0	7000	3500	0
Horizontal drilling (radials)	100	0	0	0	0	0
Lining and headworks	50	0	1500	0	0	1500
Sub-total (1994 Z$)		6516	32 500	7000	3500	32 500
Real devaluation inflator:	1.6089	CSO & RBZ				
CPI inflator:	2.7709	Idem.				
Profit inflator PWP only	1.1	Assumed				
Exchange rate, Dec 1998, US$:Z$	37.36	Idem.				
Capital costs (Dec 1998 US$)	Devaluation					
taking account of devaluation	multiplier					
Site ID by records	1.0000	0	0	0	0	0
Site ID by exploratory drilling	1.6089	0	0	0	0	0
Site ID by geophysics	1.6089	0	131	0	0	131
Pumping tests/modelling yield	1.0000	0	0	0	0	0
Vertical drilling	1.6089	855	3938	0	0	3938
Dug well or dam construction	1.3045	0	0	745	372	0
Horizontal drilling (radials)	1.6089	0	0	0	0	0
Lining and headworks	1.3045	0	160	0	0	160
Sub-total (water point)		855	4229	745	372	4229
Average yield of water (m³/day)		35.1	22.5	7.1	1.9	22.5
Pump type		hydro	hydro	bush	bucket	bush
Single pump capacity (m³/day)		11.0	11.0	18.0	3.6	11.0
Single pump cost (US$)	1.0000	800	800	245	171	490
No. of pumps to reach capacity		4	2	1	1	1
Sub-total (pumps)		3200	1600	245	171	490
Fenced area (ha)	0.032 × yield	1.12	0.72	0.23	0.06	0.00
Value of land	US$ 100/ha	112	72	23	6	0
Cost of fence	US$ 820/ha	921	590	186	50	0
Sub-total (garden)		1033	662	209	56	0
Sub-total (social development)	US$ 1000/site	1000	1000	0	0	0
Total Capital Investment		6089	7491	1199	599	4719
Recurrent costs (Dec 1998 US$)						
Depreciation: waterpoint & garden	20 yrs; 5% p.a.	94	245	48	21	211
Depreciation: pumps	10 yrs; 10% p.a.	320	160	25	17	49
Pump repairs	5% pump cost	160	80	12	9	25
Fuel	US$0.05 m⁻³	0	0	0	0	0
Labour	US$0.23 / 8hr day	1453	932	294	79	0
Total		2028	1416	378	126	285
Garden benefits (Dec 1998 US$)	Input GM/ha					
Average Gross Margin (US$/year)	2215	2488	1595	503	135	0

		Hydro-fracture existing borehole	New conv. deep borehole	Deep well av. depth 17.6 m	Family well av. depth 10.8 m	Standard domestic borehole to 55 m
Cashflow with labour costed	Year	−6089	−7491	−1199	−599	−4719
(Dec 1998 US$/year)	1	460	179	125	9	−285
recurrent costs + 13% inflation	2	520	202	141	10	−322
garden benefits + 13% inflation	3	587	228	159	11	−364
	4	664	258	180	13	−411
	5	750	291	203	15	−465
	6	848	329	239	16	−525
	7	958	372	260	19	−593
	8	1082	420	294	21	−670
	9	1223	475	332	24	−758
	10	1382	536	375	27	−856
Replace pumps	11	1561	606	424	30	−967
	12	1764	685	479	34	−1093
	13	1994	774	541	39	−1235
	14	2253	874	611	44	−1396
	15	2546	988	691	49	−1577
	16	2877	1117	780	56	−1782
	17	3251	1262	882	63	−2014
	18	3674	1426	997	71	−2276
	19	4151	1611	1126	81	−2571
	20	4691	1821	1273	91	−2906
IRR years 1-20 (% p.a.)		**16**	**5**	**21**	**1**	**negative**
Net Present Value (13%)		**2053**	**−4331**	**1010**	**−441**	**−9762**
Cashflow with labour costed	Year	−1370	-2772	3520	4119	
relative to standard domestic	1	745	463	410	294	
borehole (Dec 1998 US$/year)	2	805	487	426	295	
recurrent costs + 13% inflation	3	872	513	444	296	
gardens benefits +13% inflation	4	949	543	465	298	
	5	1035	576	488	299	
	6	1132	614	515	301	
	7	1243	657	545	304	
	8	1367	705	579	306	
	9	1508	760	617	309	
	10	1667	821	660	312	
Replace pumps	11	1846	891	709	315	
	12	2049	970	764	319	
	13	2279	1059	826	324	
	14	2538	1159	896	329	
	15	2831	1273	976	334	
	16	3162	1402	1065	341	
	17	3536	1547	1167	348	
	18	3958	1711	1282	356	
	19	4436	1896	1411	365	
	20	4976	2105	1558	376	
IRR years 1-20 (% p.a.)		**63**	**22**	**infinite**	**infinite**	
Net Present Value (13%)		**8773**	**2389**	**7730**	**6279**	
Over 20 year life assuming 13% inflation:						
Total benefits US$		201 383	129 092	40 736	10 901	N/A
Cost to community without interest but labour costed		170 237	122 132	31 833	10 778	N/A
Costs to community without interest or labour costs		52 587	46 715	8035	4410	27 784
Benefit/cost ratio (with labour costs)		1.2	1.1	1.3	1.0	N/A
Benefit/cost ratio (without labour costs)		3.8	2.8	5.1	2.5	N/A
Total cost US$/m³ of water (with labour costs)		1.01	1.13	0.93	1.18	N/A
Total cost US$/m³ of water (without labour costs)		0.31	0.43	0.24	0.48	0.53
Total cost US$/ha of irrigation (with labour costs)		7578	8481	7006	8864	N/A
Total cost US$/ha of irrigation (without labour costs)		2341	3244	1768	3626	N/A
Per capita capital cost US$ (80 families at 9.3 people)		8.18	10.07	128.92	64.45	6.34
Per capita recurrent cost US$/yr (with labour costs)		2.73	1.90	40.69	13.52	0.38
Per capita recurrent cost US$/yr (without labour costs)		0.77	0.65	9.08	5.06	0.38

Appendix 2: Pumping test examples

Introduction

Pumping test results can be analysed manually or with the aid of a computer and proprietary software package such as 'AquiferTest' or 'AquiferWin32'.

Where analysed manually, the data are plotted graphically and either curve fitting or nomograms are used to determine values for aquifer Transmissivity (*T*) and Storativity (*S*). These values are then used to estimate the safe maximum discharge (*Q*) for a given available drawdown in the well or borehole at any given time.

Where analysed by computer, the same steps of graphical presentation and curve fitting are performed automatically, and the values of T and S obtained are again used (manually) to estimate *Q*.

Where computing facilities are available, more rigorous analysis of the pumping test data is recommended, since the results are used to plan increased water supply and productive activities. A software package (BGSPT) containing programs PTFIT and PTSIM developed by Barker (1989) is now freely available on the British Geological Survey web-site (www.bgs.ac.uk/bgspt) and can be used to estimate values of *T* and *S* and to model water level behaviour under a given abstraction regime for a set period of no groundwater recharge. In this way, the safe value of *Q* likely to be sustained during drought can be projected.

The following example shows manual (graphical) analysis of a constant-rate pumping test performed using a handpump already fitted to a potentially under-utilised borehole (Test 1a in Table 4.1). Computer-generated output using programs PTFIT and PTSIM is also shown. The same procedure can be used to analyse each of the pumping tests listed in Table 4.1.

Table A2.1 Example of pumping test data

Site: Conventional deep borehole at Muzonididya Primary School
Test: 1a using the handpump fitted to the borehole
Date: 20 May 1994

Ministry of Water records:		**Pumping data**		
Borehole depth: 48 m	pump time (hrs)	1	av. pump rate $l\,s^{-1}$	0.59
Diameter: 0.15 m	start vol. (m^3)	273.614	drawdown (m)	2.96
Depth to pump inlet: 24 m	end vol. (m^3)	275.744	de-watered vol. (m^3)	0.052
Water first strike: 12 m	start wl. (mbgl)	8.8	pumped vol (m^3)	2.13
Main strike: 25 m	end wl. (mbgl)	11.76	lamda'	0.025
Rest-water level: 8.8 m				
Blowing yield: 2.1 $l\,s^{-1}$				

Manual pumping rate during test

minute	pumped volume (l)	average rate ($l\,s^{-1}$)	minute	pumped volume	average rate ($l\,s^{-1}$)
1	37	0.62	31	38	0.63
2	37	0.62	32	38	0.63
3	37	0.62	33	31	0.52
4	36	0.60	34	39	0.65
5	34	0.57	35	39	0.65
6	35	0.58	36	38	0.63
7	36	0.60	37	37	0.62
8	35	0.58	38	38	0.63
9	33	0.55	39	33	0.55
10	35	0.58	40	38	0.63
11	40	0.67	41	31	0.52
12	35	0.58	42	36	0.60
13	35	0.58	43	36	0.60
14	35	0.58	44	37	0.62
15	36	0.60	45	35	0.58
16	33	0.55	46	35	0.58
17	30	0.50	47	36	0.60
18	28	0.47	48	32	0.53
19	27	0.45	49	35	0.58
20	29	0.48	50	35	0.58
21	31	0.52	51	36	0.60
22	37	0.62	52	33	0.55
23	37	0.62	53	33	0.55
24	35	0.58	54	34	0.57
25	34	0.57	55	36	0.60
26	31	0.52	56	35	0.58
27	33	0.55	57	36	0.60
28	36	0.60	58	34	0.57
29	38	0.63	59	34	0.57
30	37	0.62	60	34	0.57
			Average		0.59

Table A2.2 Time and drawdown data

T start (mins)	T pstop (mins)	water level (mbgl)	drawdown (m)	T pstart (mins)	T pstop (mins)	water level (mbgl)	drawdown (m)
0		8.27	0.00	60.0	0.0	11.15	2.88
1		9.63	1.36	60.5	0.5	10.37	2.10
2		10.26	1.99	61.0	1.0	9.88	1.61
3		10.61	2.34	61.5	1.5	9.56	1.29
4		10.81	2.54	62.0	2.0	9.33	1.06
5		10.87	2.60	62.5	2.5	9.18	0.91
6		11.02	2.75	53.0	3.0	9.06	0.79
7		11.06	2.79	63.5	3.5	9.98	0.71
8		11.08	2.81	64.0	4.0	8.90	0.63
9		11.08	2.81	64.5	4.5	8.85	0.58
10		11.17	2.90	65.0	5.0	8.80	0.53
12		11.29	3.02	66.0	6.0	8.73	0.46
14		11.29	3.02	67.0	7.0	8.68	0.41
16		11.29	3.02	68.0	8.0	8.63	0.36
18		10.99	2.72	69.0	9.0	8.60	0.33
20		11.08	2.81	70.0	10.0	8.57	0.30
22		11.15	2.88	72.0	12.0	8.52	0.25
24		11.15	2.88	74.0	14.0	8.49	0.22
26		11.09	2.82	76.0	16.0	8.46	0.19
28		11.14	2.87	78.0	18.0	8.44	0.17
30		11.33	3.06	80.0	20.0	8.42	0.15
32		11.33	3.03	82.0	22.0	8.41	0.14
34		11.35	3.08	84.0	24.0	8.40	0.13
36		11.37	3.10	86.0	26.0	8.39	0.12
38		11.26	2.99	88.0	28.0	8.37	0.10
40		11.16	2.89	90.0	30.0	8.36	0.09
42		11.04	2.77	92.0	32.0	8.35	0.08
44		11.14	2.87	94.0	34.0	8.35	0.08
46		11.16	2.89	96.0	36.0	8.33	0.06
48		11.04	2.77	98.0	38.0	8.32	0.05
50		11.13	2.86	100.0	40.0	8.31	0.04
52		11.15	2.88	102.0	42.0	8.29	0.02
54		11.13	2.86	104.0	44.0	8.28	0.01
56		11.14	2.87	106.0	46.0	8.27	0.00
58		11.18	2.91	108.0	48.0	8.27	0.00

Curve fitting to determine T and S

Values of T and S can be determined using either standard pumping test curve matching (Theis) or straight line (Cooper and Jacob) techniques. The Cooper and Jacob equation is a simplified version of the Theis equation but has the advantage that reference tables in textbooks are not required. Hence the procedure is quite simple. For metric units, the Cooper and Jacob equation can be written thus:

$$ s_1 = \left(\frac{0.183Q}{T} \right) \times \log \left(\frac{2.25Tt}{r^2 S} \right) $$

where:

Q = well discharge rate (m^3 day^{-1})
s_1 = drawdown (m)
T = Transmissivity (m^2 day^{-1})

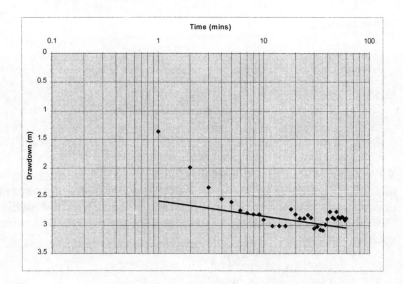

Figure A2.1   Time–drawdown analysis by the Cooper and Jacob method

S = storage coefficient (dimensionless)
t = time (days)
r = radial distance (m) (water point radius if no observation well)

The Cooper and Jacob equation plots as a straight line on semi-logarithmic paper if the limiting conditions are met. Time is plotted along the logarithmic x-axis and drawdown is plotted along the linear y-axis. Straight line plots of drawdown versus time can be produced after sufficient time has elapsed. In pumping tests with multiple observation wells, the closer wells will meet the conditions first. In our example, the water point itself is observed, and the radial distance is set equal to the borehole radius. For the Time–Drawdown method, values of T and S are calculated as follows:

$$T = \frac{0.183 \times Q}{\delta_s}$$

$$S = \frac{2.25 \times T \times t_o}{r^2}$$

where, δ_s is the change in drawdown over one logarithmic cycle, and t_o is the time value where the straight line fit of the data intersects the time axis. From the figure shown:

$\delta_s = 0.27$ m
$t_o = 0.0000000004 = 4 \times 10^{-10}$ min

In our example:

$$Q = 0.59 \text{ l/s} = 0.0209 \text{ m}^3 \text{ min}^{-1}$$
$$r = 0.075 \text{ m}$$

hence

$$T = 0.014 \text{ m}^2 \text{ min}^{-1} = 20 \text{ m}^2 \text{ day}^{-1}$$
$$S = 0.000000002 = 2 \times 10^{-9}$$

With values of T and S obtained, the Cooper and Jacob equation can be used to predict what the safe maximum yield (Q) might be for a given available drawdown (s_1) in the well or borehole at any given time (t). For metric units, the Cooper and Jacob equation shown earlier can be transposed to obtain Q thus:

$$Q = \left(\frac{s_1 \times T}{0.183} \right) \div \log \left(\frac{2.25Tt}{r^2 S} \right)$$

Both the standard Theis and Cooper and Jacob equations assume the aquifer to be homogenous, which in practice of course it never is. In basement aquifers, therefore, it is unrealistic to extend the use of these equations into the bedrock where fissure flow dominates. For most practical purposes the base of the weathered zone or regolith should be taken as the base of the aquifer. The maximum available drawdown then becomes the depth of saturated regolith. However, such a drawdown is unrealistic since there would be no aquifer left if water levels were lowered to this point. Traditionally, the maximum available drawdown is taken to be two-thirds of the total depth of saturated aquifer, or the inlet depth of the pump if this is less.

In our example, the maximum depth of regolith in this environment is assumed to be of the order of 30 m, the measured rest water level is 9 m, and the length of dry season over which the water point must survive without recharge is set at 240 days. Hence:

$$s_1 = 2/3 \times (30\text{-}9) = 14 \text{ m}$$
$$t = 240 \text{ days}$$
$$r = 0.075 \text{ m}$$
$$T = 20 \text{ m}^2 \text{ day}^{-1}$$
$$S = 2 \times 10^{-9}$$

giving

$$Q = \left(\frac{14 \times 20}{0.183} \right) \div \log \left(\frac{2.25 \times 20 \times 240}{0.075 \times 0.075 \times 2 \times 10^{-9}} \right)$$

$$= 102 \text{ m}^3 \text{ day}^{-1}$$

It must be stressed that both the standard Theis and Cooper and Jacob equations are based on many assumptions, most of which are not met in reality in hard rock aquifers. Obtaining values of T and S using data from the pumped well rather than an observation well is also less than ideal. The result, therefore, should only be viewed as an 'order of magnitude' value and not a precise prediction.

Results using programs PTFIT and PTSIM

Programs PTFIT and PTSIM available from www.bgs.ac.uk/bgspt will run on any personal or mainframe computer. PTFIT models drawdown and recovery data to determine aquifer properties T and S. PTSIM uses these properties to simulate water-level response to a daily pumping regime, set in our example at 15 m^3 day^{-1} (1500 l hr^{-1} from 06:00 to 11:00 in the morning and from 13:00 to 18:00 in the afternoon) repeated for an extended dry period of 240 days. The drawdown at the end of this period is then scaled up or down to equal the available drawdown in the water point to define the maximum abstraction rate possible.

Analysis of the constant-rate test data shown above using PTFIT gives aquifer properties T=32 m^2 day^{-1} and S=2 × 10^{-6}, somewhat different to the values obtained manually using the simple Cooper and Jacob method. Figure A2.2 shows a good fit between observed test data and the modelled response suggesting a high level of confidence in the result. These aquifer properties, used in PTSIM, predict the maximum yield that could be sustained over an extended dry period of 240 days to be of the order 200 m^3 day^{-1}, confirming again that the potential safe yield is far higher than present use.

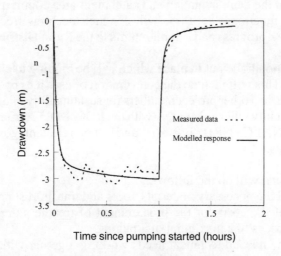

Figure A2.2 Modelled response using programme PTFIT

Appendix 3: Social contracts and constitutions

Social Contract: Mbiri Dam Rehabilitation Project

The name of the project shall be Mbiri Dam Rehabilitation. Mbiri dam is in Shurugwi District Ward 14. It lies 20 km from Tongogara Rural District Council, along Tongogara Nhema road and 800 m from Jobolingo Township.

AGREEMENT :
The community will do the following tasks:
1. Provide all the needed labour on the execution of the project which includes:
2. Catchment Area Rehabilitation.
3. Fencing off of the dam.
4. Repairs / upgrading of the dam.
5. Irrigation activities.
6. Marketing of produce.
7. Water harvesting.
8. Safely keep the project materials and tools for use at the agreed project only.
9. Ensure that the dam, irrigation and catchment area committees are in place and functional i.e. these will conduct meetings frequently to review work progress and submit reports to the Rural District Council and CARE.
10. A constitution to be put in place which will be binding to all affected members. This will include the contribution of user fees by irrigation participants and other project members for sustainability of the project.
11. The community will work closely with the RDC, DDF, Agritex, CARE, MOH, MNAEC, NRB, Forestry and other relevant government departments.

CARE International will do the following:
1. Provide all the necessary materials, tools and transport support.
2. Pay half of the costs for the procurement of treadle pumps for the project gardens to a maximum of 4 pumps.
3. Provide any needed technical needs working together with Agritex, NRB, DDF, Forestry Commission and any other organization.

4. Monitor the project and co-ordinate in providing necessary training.
5. The following signatures on this form are a symbol of agreement between CARE and the Project Community.

Mbiri Dam Committee:

Chairperson:	Wilson Chikwekwete	——————
Vice:	Venencia Machacha	——————
Secretary:	Temba Ziyambi	——————
Vice:	Susan Wadyajena	——————
Treasurer:	T. Msindo	——————
Comm. Members:	Clifford Machacha	——————
	Pauline Chinyama (Mrs)	——————
Security:	Leonard Munemo	——————

Date: 14 February 1997

Zaka Rural District Council
Private Bag 31Z, Zaka
17 January 1994

Social Contract for Construction, Ownership, Maintenance and Repair of a Productive Water Point in your Home Area

To the Community:
Mushungwa & Chigarera Village Development Committees (VIDCOs)
Ward 27, Zaka District

Subsequent to our meeting of 6th December, at which you confirmed your interest in implementing a collector well and 0.56 ha community garden, the following information is provided for your consideration, and the following conditions outlined for your approval:

Ownership
○ The productive water point and garden once complete will belong entirely to you the community.
○ This project is designed to allow you the community to help yourselves, by contributing fully to the construction, and by subsequently managing and maintaining the well and garden yourselves.
○ The well and garden will belong to you entirely, and not in any way to the visiting project staff. These staff will only take part during construction, helping you to create your own project. Thereafter, these people will visit only to see how things are going. Your help in providing information during these visits will be appreciated.

211

Land allocation
- ○ Construction can begin after you the community, in full consultation, have identified the land to be used henceforth for this project, and have agreed compensation in writing with the present landowner.

Our contributions
- ○ During construction, project staff will provide:
  A resident foreman to supervise construction
  Materials for the well including steel well lining and cement
  Digging tools e.g. spades, picks, boots, hats
  Machinery and fuel e.g. compressor, jackhammer, diesel
  Steel fence and gate for a garden 75 m × 75 m square
  Crushed stone filter for the well
  Two Zimbabwe 'B' type handpumps for the well
  Maintenance tools, a rope, a manual and a gantry above the well to allow you to undertake pump repairs.

Your community contributions
- ○ Five strong men per day to dig the well. Working under direction of the foreman, each group of five men should work for six days before changing to the next group. Construction will take approximately 12 weeks.
- ○ Security for equipment at the site. It is probably best if you organize a night watchman system.
- ○ Labour to erect the garden fence. Working under direction of the foreman, many hands will make light work.
- ○ Sieved river sand, gravel, stones, rocks. Also make approximately 250 bricks. A local builder will be helpful.
- ○ At least 10 local men and women who live permanently in the area to be trained in pump maintenance and repair.

Your community responsibility for all management, maintenance, repair and improvements
- ○ As outlined, all benefits from the project will belong entirely to you the community. However, as the sole owners of the project you will be entirely responsible for all management decisions and, most importantly, for all upkeep, maintenance and repair that will be necessary in future. This includes maintenance and repair of the pumps fitted to the well. Once construction is complete and you have received training, project staff can provide no further assistance or materials. You should ensure that, as a community, you are able to repair or can pay others to repair, your project. You are responsible for any improvements that you might wish to make in the future.

Suggestions to help the project be a success
- ○ The community should elect who they wish to be responsible for the project as described. A series of community meetings to discuss membership will help. To avoid problems in future, it will be necessary

to ensure that all members of the community have a chance to join the project at the beginning. Once membership is agreed, a small fee might be collected from each person (perhaps $10-$20). This money can be used to form an initial project fund to buy seeds and sprays needed for the first cropping season.

Agreement
- O If, on reading this letter, you as a community still wish to construct a collector well and 0.56 ha community garden in your home area, if you agree to provide the community contributions outlined, and if you agree to be fully responsible for all management, maintenance and repair of the project once completed, please select four official representatives to sign below, printing their names and positions within the community.

Signature: ....
Name: ....
Position: .... Date: ....

Constitution: Mbilashava Dam Rehabilitation project

1. <u>Name</u>: Mbilashava Dam Rehabilitation and Irrigation

2. <u>Location</u> : 18 km north of Zvishavane off the Gweru road between Mbilashava and Ngomeyebani VIDCOs.

3. Objectives
- O To establish a community system to lead the work.
- O To erect silt traps.
- O To standardize all fields in the catchment area.
- O To abate stream bank cultivation.
- O To prevent silt from all sources in the catchment area from coming to the dam.
- O To protect the dam reservoir and wall and improve the gardens.
- O To seek technical advice from all government departments and NGOs.

4. <u>Area of Influence</u>
Mbilashava and Ngomeyebani VIDCOs.

5. <u>Membership</u>
Free of charge and voluntary to all residents of Mbilashava and Ngomeyebani VIDCOs and any other interested parties from neighbouring Vukuso ward.

All with a garden at the dam, owning cattle, goats, donkeys or sheep, with a field in the catchment area or who use dam water in any way shall be expected to participate.

6. Rights of Community
Decision making at meetings.

7. Obligation of Members
Abide by constitution and direction given by chosen committee and extension staff.

8. Termination of Membership
When one has left the area of influence or upon project completion.

9. Voting
Each member is allowed to vote.

10. Quorum
When there is sufficient representation (half the participants) from each village.

11. General Meeting
Shall be held once every month.

12. Management Committees, (Dam, catchment area and irrigation)
The dam committee shall serve as the mother body with absolute control over the whole project and shall have a catchment area committee in terms of work in the catchment area and an irrigation committee in terms of gardens administration.

13. Term of Office of Management Committees
One year, after which new elections are held at an annual general meeting. If one is found defective on three occasions they will be asked to step down.

14. Powers of the Management Committees
To cause all members of the project to abide by the set bye-laws, to set up sub-committees to carry out certain activities, to inspect the catchment area, dam and gardens, to collect subscriptions (user fees) from garden participants.

15. Duties of the Management Committees
To administer the day-to-day work at the dam sites.
To monitor water use and activities at the dam site and gardens.
To maintain proper account of all monies, tools and materials at project site and project registers and a log book.

16. Duties
Chairperson: To head all project activities, keep tools and materials.
Secretary: Maintain an up-to-date record of all records, documents and registers of the dam. Keep inventory of the dam property. Sign and conduct the correspondence of the dam. Summon members to attend meetings. Keep and take minutes at all meetings.
Treasurer: Handling all monies coming to the dam or irrigation committee. Keep all accounts and books of the project.

17. Banking
All monies shall be banked and books shall be kept (books of accounts). The chairperson, secretary and treasurer shall be signatories. Any two will sign.

18. Registers and Records
These shall be kept by secretaries of the dam, irrigation and catchment area committees and should be used whenever at work.

19. Settlement of Disputes
Shall be done through the dam committee with the Councillor sitting in failure to which the matter will be referred to the District Administrator.

Irrigation Committee:

Chairperson:	I. Machacha	_____
Vice:	T. Ziyambi	_____
Secretary:	J. Musvire	_____
Vice:	S. Mangaliso	_____
Treasurer	M. Kwashira	_____
Comm. members	L. Mashonganyika	_____
	E. Chikwekwete	_____

Date: 14 February 1997

Appendix 4: Groundwater development and the concept of safe yield

Continued increase in the use of groundwater makes more pressing the question as to the extent of the reserves and the advisability and methods of regulation. Proper regulation is conditional upon the ability to forecast with some degree of accuracy the future of water levels in wells in a given area. Perhaps the most common misconception in groundwater management is that a water budget of an area determines the magnitude of possible groundwater development or, to put it another way, the idea that pumping within a groundwater basin shall not exceed the recharge. Theis (1940) first addressed the subject:

> All groundwater is in the process of movement through the aquifer from a place of recharge to a place of discharge. Changes in the quantity of water in the aquifer, with accompanying changes in water level, occur as the result of temporary imbalance between recharge and discharge by natural processes, but such fluctuations balance each other over time. Under natural conditions, therefore, previous to development by wells, aquifers are in a state of approximate dynamic equilibrium. Discharge by wells is thus a new discharge superimposed upon a previously stable system, and it must be balanced by an increase in the recharge of the aquifer, or by a decrease in the old natural discharge, or by a loss of storage in the aquifer, or by a combination of these.

Theis outlines that there are three essential factors controlling the action of an aquifer: (1) the distance to, and character of, the recharge; (2) the distance to the locality of natural discharge; (3) the character of the cone of depression of the water table around a pumping well which depends on the coefficients of transmissivity and storage of the given aquifer. He concluded with six key points for the ideal development of any aquifer from the standpoint of groundwater conservation and maximum utilization of the supply:

1. All water discharged by wells is balanced by a loss of water somewhere.

216

2. This loss is always to some extent and in many cases largely from storage in the aquifer. Some groundwater is always mined. The reservoir from which the water is taken is in effect bounded by time and by the structure of the aquifer as well as by material boundaries. The amount of water removed from any area is proportional to the drawdown, which in turn is proportional to the rate of pumping. Therefore, too great a concentration of pumping in any area is to be discouraged and uniform areal distribution of development over the area where the water is shallow should be encouraged.

3. After sufficient time has elapsed for the cone of depression to reach the area of recharge, further discharge by wells will be made up, at least in part, by an increase in the recharge *if* previously there has been rejected recharge. If the recharge was previously rejected through transpiration from non-beneficial vegetation, no economic loss is suffered. If the recharge was rejected through springs or as surface waters, rights to these surface waters may be injured.

4. Again, after sufficient time has elapsed for the cone to reach the areas of natural discharge, further discharge by wells will be made up in part by a diminution in the natural discharge. If this natural discharge fed surface streams, prior rights to the surface water may be injured.

5. In large non-artesian aquifers, where pumping is done at great distances from the localities of intake or outlet, the effects of each well are for a considerable time confined to a rather small radius and the water is taken from storage in the vicinity of the well. Hence these large groundwater bodies cannot be considered a unit in utilizing the groundwater. Proper conservation measures will consider such large aquifers to be made up of smaller units, and will attempt to limit the development in each unit. Such procedure would also be advisable, although not as necessary, in a large artesian aquifer.

6. The ideal development of any aquifer from the standpoint of the maximum utilization of the supply would follow these points:

a) The wells should be placed as close as possible to areas of rejected recharge or natural discharge where groundwater is being lost by evaporation or transpiration by non-productive vegetation, or where the surface water fed by, or rejected by, the groundwater is not used. By so doing this lost water would be utilized by the wells with a minimum lowering of the water level in the aquifer.

b) In areas remote from zones of natural discharge or rejected recharge, the wells should be spaced as uniformly as possible throughout the available area. By so doing the lowering of the water level in any one place would be held to a minimum and hence the life of the development would be extended.

c)   The amount of pumping in any one locality would be limited. For non-artesian aquifers with a comparatively small areal extent and for most artesian aquifers, there is a perennial safe yield equivalent to the amount of rejected recharge and natural discharge it is feasible to utilize. If this amount is not exceeded, the water levels will finally reach an equilibrium stage. If it is exceeded, water levels will continue to decline.

d)   In localities developing water from non-artesian aquifers and remote from areas of rejected recharge or natural discharge, the condition of equilibrium connoted by the concept of perennial safe yield may never be reached in the predictable future and the water used may all be taken from storage. If pumping in such a locality is at a rate that will result in the course of 10 years in a lowering of water level to a depth from which it is not feasible to pump, pumping at half this rate would not cause the same lowering in 100 years. Provided there is no interference by pumping from other wells, in the long run much more water could be taken from the aquifer at less expense.

Glossary of technical terms

Acid treatment   Using acid to remove deposits from the face of a borehole or well to increase the yield. Synonym **acidification.**

Alluvial   Eroded material deposited by flowing water.

Annular space   The space or cavity between the outside of the borehole or well casing and the surrounding ground.

Aquiclude   A body of poorly permeable rock that is capable of slowly absorbing water from, and releasing water to, an aquifer. It does not transmit groundwater rapidly enough by itself to directly supply a borehole, well or spring. Synonym **aquitard**.

Aquifer   A water-bearing geological formation capable of yielding groundwater in useful amounts to wells and boreholes. From the Latin words, *aqua* (water) and *ferre* (to carry). An aquifer literally carries water – underground. A stratum which contains intergranular interstices and or a system of interconnected fractures capable of transmitting groundwater rapidly enough to directly supply a borehole, well or spring. An aquifer may be only a few metres thick or hundreds of metres thick, it may lie just under the surface or hundreds of metres down, it may underlie a few hectares of land or thousands of square kilometres. It is generally bound by either an aquiclude (aquitard) or aquifuge.

Aquifuge   A rock which contains no interconnected openings and therefore neither absorbs not transmits water.

Artesian aquifer   An aquifer in which the groundwater is under pressure and confined beneath impermeable rocks; the water level rises above the top of the aquifer and sometimes may flow out at the surface when a borehole or well penetrates an artesian aquifer.

Base flow   That part of stream flow which is contributed by effluent groundwater. It sustains the stream during dry periods.

Bedrock   Solid rock underlying the weathered rock or regolith.

Borehole   A type of water well mechanically drilled and characterized by a relatively small diameter (100-450 mm) and large depth, typically 60-70 m. Synonym **tubewell.**

Brecciation   Rock debris and weathering within a fracture or fissure.

Capillarity   The action by which water is drawn up (or depressed) in small interstices as a result of surface tension.

Casing   Steel or plastic pipe used to support the sides of a well or borehole. It may be slotted with saw cuts to allow water to enter and usually has screw joints. Synonym **lining**.

Catchment   In the physical sense is an area from which rainfall drains into a stream or river or dam or lake. Synonym **watershed**, although the Oxford English Dictionary defines watershed as a line of high land where streams on one side flow into one river and streams on the other side flow into another. While the terms differ in meaning, both provide an appropriate focus for water resource management, and the process taken and results obtained do not differ.

Centrifugal pump A pump using centrifugal force to move water.

Chlorination A method of disinfecting water to reduce microbial contamination using chlorine gas or a solution of sodium hypochlorite or bleaching powder, all of which contain chlorine.

Cone of depression The inverted cone-shaped depression of the water table around a pumping well or borehole. The shape and extent of the cone of depression depends on the rate and duration of pumping, and the hydraulic characteristics of the aquifer.

Crack A partial or incomplete fracture.

Collector well A large-diameter well with horizontal boreholes drilled radially outwards from the base. Synonyms **Rannay-type well** and **dug-cum-bore well**

Confined groundwater Groundwater under pressure significantly greater than that of the atmosphere and whose upper surface is the bottom of a layer of distinctly lower permeability than the material in which the water occurs. See also **Artesian aquifer**.

Crystalline basement Rock of igneous or metamorphic origin primarily of Precambrian age and granitic or gneissose in type.

Deep well A hand-dug well that has been deepened beyond the zone of alteration and into fresh bedrock, usually with the aid of dynamite and or mechanical digging tools.

Developing a well Increasing the yield of a newly constructed well or borehole by flushing or swabbing a plunger up and down to remove sand, silt and mud which is adhering to the well face.

Discharge The volume of water flowing in a stream or from a spring or from a pipe in a unit of time e.g. litres per second. It is also the delivery rate of a pump.

Divining A mystical method of locating groundwater or an auspicious place to construct a well using a hazel twig, brass rods, an empty bottle or other device. When submitted to impartial scientific tests divining has not been successful. Synonym **dowsing**.

Drawdown   The difference between the observed water level during pumping and the water level before pumping commenced.

Effective rainfall Rainfall minus interception losses minus storm runoff, in other words, that part of the rainfall that wets the soil.

Evapotranspiration Loss of water from a land area through transpiration of plants and evaporation from the soil.

Fissures Natural cracks in rocks, which greatly enhance groundwater movement.

Fracture Any break in a rock whether or not it causes displacement, owing to mechanical failure by stress. Fracture includes cracks, joints and faults.

Geohydrology The branch of hydrology dealing with subsurface water i.e. water in both the saturated and unsaturated zones. Synonym **hydrogeology**.

Gneissosity The coarse textural banding of the constituent minerals in a metamorphic rock, commonly a gneiss, into alternating light (silicic) and darker-coloured (mafic) layers.

Gravel pack Gravel or coarse sand which is inserted around the slotted screen in a borehole or well to prevent fine-grained aquifer material from being drawn into the borehole or well.

Grike A vertical fissure developed along a joint in limestone or dolomite through solution.

Groundwater Water which is contained in saturated rock. It flows into boreholes and wells or debouches as springs. Synonym **underground** or **subterranean water**.

Groundwater level See **rest-water level**.

Grout A cement slurry used to fill the annular space between the ground and the lining of a well or borehole so that unwanted surface water cannot get into the well or borehole.

Hand-dug well A shallow water well, typically 0.8-1.2 m in diameter, completed in the regolith by manual labour. Synonyms **open well** or **dug well** or **family well**.

Hard rock A compact rock that lacks primary porosity.

Head The potential energy of water due to its height above a given level, expressed in metres. It is used for the height to which pumps must lift water and also the pressure in a distribution system due to the presence of a reservoir.

Head losses The pressure losses in an aquifer, well or pipe system expressed as a head of water.

Headworks The structure and pipework constructed at the top of a well or borehole.

Hydraulic conductivity Rate of flow of water through a unit cross-section of soil or rock under unit hydraulic gradient. It depends on many factors, such as porosity, size of openings, interconnection between openings, rounding and sorting of grains, etc.

Hydraulic gradient The difference in hydraulic head at two points within a water system such as a pipework, stream or aquifer, divided by the distance between the two points measured along the flow path. Flow can take place only if a hydraulic gradient exists.

Hydrogeology The geology of groundwater. Synonym **Geohydrology**.

Hydrologic cycle The series of interlinked processes which cause water to be circulated from the oceans to the atmosphere to the ground as precipitation and returned to the oceans as river and groundwater flow.

Hydrology The science that deals with continental water on and under the Earth's surface.

Igneous rock Formed as molten rock cools and hardens.

Impermeable A description of rock, soil or other material that will not transmit water.

Interception The process by which water from precipitation is caught and stored on plant surfaces and eventually returned to the atmosphere without having reached the ground.

Interflow That part of precipitation which flows across the surface of an impermeable layer at shallow depth within the soil profile.

Karst A type of topography that is formed over limestone and dolomite by solution and that is characterized by closed depressions or sinkholes, caves and underground drainage.

Large-diameter well A well of internal diameter 2–10 metres constructed manually, usually with the aid of a de-watering pump and mechanical tools to allow completion below the water table to the depth of fresh rock. Synonym **agro-well**.

Lithology The description of rocks on the basis of colour, structures, mineralogical composition and grain size.

Metamorphic rocks Igneous rocks changed in texture or structure by heat and pressure.

Natural groundwater recession That part of groundwater recharge which discharges or is lost naturally as transpiration by trees and other deep-rooted vegetation, lateral flow, base flow, and deep percolation. Synonym **natural groundwater discharge.**

O&M Operation and maintenance.

Outcrop That part of a rock which is exposed at the surface.

221

Perched water table An area of saturated aquifer retained above the main water table by a small area of impermeable rock.

Percolation Water moving downwards under the influence of gravity.

Percussion drilling A method of constructing a borehole using cutting tools that are suspended on a steel rope. The drilling action gives repeated blows to the rock at the bottom of the borehole. The broken rock is then removed using a bailer.

Permeability The capacity of rock and soil to transmit water. See also **hydraulic conductivity**

pH value A means of expressing the acidity or alkalinity of a solution. The neutral point is pH value 7.0, with acids having lower values and alkalis having higher values.

Piezometer A small-diameter borehole or tube specially constructed for the measurement of rest water level or hydraulic head at a specific depth within an aquifer system.

Pit latrine A form of lavatory where the faeces accumulate in an underground pit.

Pore spaces Microscopic spaces or interstices between individual grains in a rock. Good aquifers have interconnecting pores.

Porosity The proportion of open voids in a unit volume of rock. The porosity of the regolith has been estimated at 20-40%.

Primary porosity Interstices that were made at the time of the formation of the sedimentary deposit or rock that contains them.

Pumping test A method of testing a well or borehole to establish the reliable yield and to find out if a new well affects other wells and springs. The test consists of pumping at a known rate for a sustained period while monitoring the discharge rate and water levels in the pumping well, and in surrounding wells and local springs where available.

Recharge Water which percolates to increase the quantity of groundwater stored in an aquifer. Synonym **replenishment.**

Recovery The difference between the observed water level during the recovery period after cessation of pumping and the water level measured immediately before pumping stopped.

Regolith The mantle of fragmental and loose material of residual and/or transported origin, comprising rock debris, alluvium, aeolian deposits, soil and *in situ* weathered rock. It overlies or covers more solid rock, so-called bed rock.

Rest-water level The water level in a borehole or well which is unaffected by pumping. It represents the local level of the water table. Synonym **static water level** or **groundwater level** or **standing water level**.

Rotary drilling A rapid method of drilling boreholes. It uses cutting tools suspended on screw jointed rods which are rotated to produce the cutting action. The broken rock is removed by the circulation of a fluid in the borehole. Fluids used include air, water, special drilling foam and various types of mud.

Rotary pump A pump which has a rotating element (rotor) shaped into a slow spiral which is rotated inside a flexible moulded sheath (stator) to produce a pumping action capable of raising large volumes of water at low pressures. Synonym **progressive cavity pump**.

Runoff That part of precipitation which flows across the ground surface into streams.

Safe yield Traditionally defined as the attainment of a long-term balance between the amount of groundwater withdrawn and the amount of recharge, this concept is now discredited since it ignores natural discharge from the system and is not sustainable yield (Chapter 7).

In the context of pumping test appraisal of wells and boreholes (Chapters 4 and 5) safe yield is used here to define the reliable yield that will be sustained by the water point during a set period of no recharge (e.g. 240 days) based on projections of drawdown over this period. This is different to the maximum yield at which the water point can be pumped at any point in time, determined by standard short-duration pumping tests, which varies as seasonal fluctuations in water table change the maximum drawdown available.

Salination   The build up of mineral salts in the soil caused by the evaporation of irrigation waters.

Salinity   The total concentration of dissolved minerals in water.

Saprock   Slightly weathered rock, often having better permeability.

Saprolite   *In situ* thoroughly decomposed rock.

Saturation, Zone of   The zone below the water table in which all interstices are filled with groundwater.

Screening   A type of borehole casing specially designed to allow the efficient entry of water into the borehole.

Secondary porosity   Interstices that were made by processes that affected the rocks after they were formed.

Sedimentary rock   Eroded material laid down (deposited) by water, ice or wind.

Small dam   Low earth dam across an ephemeral stream to check and store surface runoff and promote groundwater recharge. Synonym **percolation tank** or **infiltration tank** or **check dam.**

Specific capacity   The ratio of the discharge to the drawdown it produces, measured inside the borehole or well (l/min/m of drawdown).

Specific yield   The ratio of the volume of water that a unit volume of an unconfined aquifer will release when drained by gravity. Typical regolith figures are 0.01–0.1. Synonym **effective porosity**.

Spring   A natural outflow of groundwater at the ground surface.

Steening   The brick or stone lining to a hand-dug well. Synonym **brick caisson**.

Storage coefficient   The volume of water held in confined and semi-confined aquifers which can be released by gravity drainage. It is expressed as a percentage of the total volume of the aquifer. The values generally lie between 10^{-5} and 10^{-3} (fraction of volume). Synonym **storage capacity**.

Storativity   The volume of water an aquifer releases or takes into storage per unit surface area per unit change in head. In an unconfined aquifer, it is normally referred to as **specific yield**. In a confined aquifer, it may be referred to as **storage coefficient**.

Stratigraphy   The description of strata – sedimentary, igneous and metamorphic – in terms of lithologic composition, fossil content, age, origin, history.

Submersible pump   A pump which is designed to operate under water. Usually, these are electric centrifugal pumps and have the electric motor enclosed in a waterproof casing. Sometimes these pumps are driven by rods which are rotated by a surface motor.

Test well   A borehole drilled to provide information on groundwater quality and geologic conditions and to test an aquifer by means of pumping tests. Synonyms **test hole** or **exploratory hole** or **exploration borehole** or **investigation borehole** or **pilot borehole**.

223

Transmissivity   The capacity of rock to transmit water. It is the product of hydraulic conductivity and aquifer thickness (m^2 day^{-1}).

Unconfined aquifer   Groundwater of which the upper surface is at atmospheric pressure, or in other words, its upper surface is the water table.

Unsaturated zone   The zone between the land surface and the water table in which interstices contain air or gases generally under atmospheric pressure as well as water under pressure less than that of the atmosphere. Synonym **zone of aeration.**

VLOM   Village-level operation and maintenance.

Water table   The surface of saturation in an unconfined aquifer at which the pressure of the water is equal to that of the atmosphere.

Wind pump   A form of piston or rotary pump which is powered by a vane rotated by the wind.

Index

227

229